DISPOSABLE CITIES

To my parents

Disposable Cities

Garbage, Governance and Sustainable Development in Urban Africa

GARTH ANDREW MYERS
University of Kansas, USA

Routledge
Taylor & Francis Group

LONDON AND NEW YORK

First published 2005 by Ashgate Publishing

2 Park Square, Milton Park, Abingdon, Oxon OX14 4RN
711 Third Avenue, New York, NY 10017, USA

Routledge is an imprint of the Taylor & Francis Group, an informa business

First issued in paperback 2016

British Library Cataloguing in Publication Data
Myers, Garth Andrew
  Disposable cities : garbage, governance and sustainable
  development in urban Africa. - (Re-materialising cultural
  geography)
  1. City planning - Environmental aspects - Africa,
  Sub-Saharan 2. Community development, Urban - Africa,
  Sub-Saharan 3. Municipal government - Africa, Sub-Saharan
  4. Sustainable development - Africa, Sub-Saharan 5. Urban
  ecology - Africa, Sub-Saharan 6. Political ecology - Africa,
  Sub-Saharan 7. Refuse and refuse disposal - Africa,
  Sub-Saharan
  I. Title
  307.1'216'0967

Library of Congress Cataloging-in-Publication Data
Myers, Garth Andrew.
  Disposable cities : garbage, governance and sustainable development in urban Africa
/ by Garth Andrew Myers.
      p. cm. -- (Re-materialising cultural geography)
  Includes bibliographical references and index.
  ISBN 0-7546-4374-3
  1. Urbanization--Africa. 2. Sustainable development--Africa. 3. Refuse and refuse
disposal--Africa. 4. Urban ecology--Africa. I. Title. II. Series.

  HT384.A35M94 2005
  307.76'096--dc22

                                                                            2005014421
  ISBN 978-0-7546-4374-6 (hbk)
  ISBN 978-1-138-26676-6 (pbk)

Transferred to Digital Printing in 2014

Contents

List of Figures

List of Tables

Foreword

To call something disposable is to suggest that it is made to be discarded or disregarded. Kevin Bales (1999), in his book, *Disposable People*, wrote about both new and persistent forms of slavery that remain invisible for much of the western world. By using the title *Disposable Cities*, I am drawing attention to two matters at once, the first of which is at least tangentially related to the Bales book. That is the manner in which the cities of Sub-Saharan Africa have come to be discarded and disregarded, not unlike the slaves that Bales interviewed for his book, so that they are invisible for western media and under-represented in academic urban studies. The second is the contradictory outcomes of the processes of sustainable development planning and solid waste management I am studying here. I focus on a program developed by the United Nations and funded by the development agencies of many wealthy countries that ostensibly is designed to help these cities collect and dispose of their garbage, among other things. Captivated by the rhetoric of this program, which seeks to build more inclusive, participatory, decentralized, and democratic planning systems that both aid poor urban neighborhoods economically and improve their environments, I set out to understand how it had, and had not, worked in a set of cities – Dar es Salaam, Zanzibar, and Lusaka – where I have spent a lot of time over the last decade.

There are a number of recent books that seek to assess the changes afoot in the planning regimes and processes of other African cities. Some, like Emmanuel Mutale's (2004) book on Kitwe, *The Management of Urban Development in Zambia*, are more managerial or technical in orientation. Although I certainly discuss programmatic and technical planning dimensions, I have deliberately broadened the canvas. This is because so many narrative lines intersect in the planning stories of the cities I examine. To attempt to rein them in, however, I structure the book around four themes in three case studies. The themes are basically economic, environmental, political, and cultural in orientation, respectively. I consider these four themes – which I label neoliberalism, sustainable development, good governance, and the politics of cultural difference – the principle dynamics at work within and around the United Nations Sustainable Cities Program's operations in the three cities I examine. With this combination of breadth and structure, I follow in the footsteps of Maliq Simone's (2004) book, *For the City Yet to Come: Changing African Life in Four Cities*. Although I have a specific interest in solid waste management, I share in Simone's hopefulness, expressed in his title, that the narrative I have developed can provide a snapshot of cities in motion and a constructive critique of the means by which everyday life is governed.

Although I address economic, environmental, and political issues in this book alongside and intertwined with cultural ones, it is published in a series that aims at rematerializing *cultural* geography. To me, to interweave what might still be called the political economy of the urban environment with cultural geography is to rematerialize cultural geography. After all, what is more material than garbage? The decades since formal independence (which came between 1957 and 1968 for most countries in the region) and the conditions that come with that time are my main concern here. Yet I share with Mutale (2004, p. 17) the crucial contention that even in critiquing "contemporary policy failures in urban management, the historical roots of existing... management" must be "properly understood." Hence each case study chapter has some historical geography in it as well.

I have organized the book into six chapters. The first introduces the four themes and the broad dynamic of urbanization in sub-Saharan Africa, leading toward an articulation of my approach to them. The second introduces the institutional landscape behind the UN Sustainable Cities Program, while discussing both that program's framework and objectives and the intentions of the donors who back it. Chapters three though five are the heart of the book, the stories of the program's operations and outcomes in Dar es Salaam, Zanzibar, and Lusaka, respectively. The final chapter offers both a summary of my conclusions and a postscript that serves as a commentary on where this book fits in and converses with the study of African cultural and development geography.

Acknowledgments

The research for this book was made possible by the support of the Fulbright Africa Regional Research Program, the US Department of State Bureau of Educational and Cultural Affairs University Affiliations Program, and the Humanities General Research Fund of the University of Kansas. I would like to acknowledge the support, advice, and counsel of my colleagues and friends in Dar es Salaam, Lusaka, and Zanzibar in conducting the research. Many thanks to Dr. Camillus Sawio, Dr. Salome Misana, and Dr. Phillip Mwanakuzi of the Geography Department of the University of Dar es Salaam, as well as Dr. Tumsifu Nnkya, Dr. Francos Halla, Dr. Bituro Majani, and Dr. Fred Lerise of the University College of Lands and Architectural Studies in Dar for talking with me about the project. Dr. Lerise deserves special mention for his welcoming demeanor and honest advice, and for his care in reading drafts of chapters. Officials in the Dar es Salaam City Council and Kinondoni Municipal Council also proved extremely welcoming and forthcoming, especially Asteria Mlambo, Elias Chinamo, Martin Kitilla, and Joash Nyitambe. Walter and Frida Bgoya, Zuberi Waziri, Hermes Mutagwaba, Judith Mhina, and Jeff Galvin showed great kindness to my family in Dar as well. Makame Ali Muhajir, Ali Hasan Ali, Parmukh and Jussy Singh, and their families provided the same in Zanzibar. I grieve for the tragic loss of Jaswant Singh in April 2004. My research in Zanzibar also benefited from the wisdom of many people in interviews and discussions, including Sheha Mjaja Juma, Ramadhan Keis, Sigrid Pessel, Beatrice Trenkmann, Che Guevara Mwakanjuki, Ali Khalil Mirza, Wolfgang Scholz, and, of course, Muhajir. I am very grateful to Kassim Omar Juma, Yunis, Kasu, Bim, Bi Rahma, and the whole team of workers in Mkele, as well as the ordinary Mkele residents I spoke with, for allowing me to participate in the solid waste program as it began in 2003. In Lusaka, I must begin with my deep gratitude to my colleagues at the University of Zambia. Wilma Nchito collaborated with me in many aspects of the fieldwork in Ng'ombe and Kamanga, and I truly could not have done any of this work without her advice and analysis. Similarly, Dr. Mark Mulenga, Dr. Greata Banda, Godfrey Hampwaye, Iwake Masialeti, Evaristo Kapungwe, and John Volk of the Geography department and Simasiku Simasiku, Simon Nkemba, Dr. Daniel Nkhuwa, Dr. Lordwell Witika, and above all Dr. Imasiku Nyambe of the School of Mines gave of their time and insights. Mweetwa Mudenda assisted in a number of interviews and focus groups. The members, officers, and customers of the Samalila Ukhondo, Zaninge, and Kwawama waste groups in Kamanga and Ng'ombe, particularly White Phiri, Peter Tembo, and Ruth Mundia facilitated many fruitful discussions and warm receptions in their respective neighborhoods.

As the work has progressed, I have shared different written portions and basic concepts with colleagues, friends, and students. I have presented various portions in different conferences, and particularly wish to thank Claire Mercer, Ben Page, Felix Driver, Matthew Gandy, Alan Gilbert, Richard Dennis, Richard Munton, James Sidaway, Giles Mohan, Jenny Robinson, Bill Freund, Martin Murray, Maliq Simone, Karen Hansen, Rick Schroeder, Richard Harris, Debby Potts, Giacomo Macola, Kevin Blake, and Jeff Smith for making those opportunities possible. Ben took time from his own hectic research schedule to arrange my two-week stay at University College London in 2005 where I completed the last pieces of the manuscript. Jenny has read chapters and as always offered great encouragement throughout. At the University of Kansas, my students Ang Gray, Sarah Smiley, Iwake Masialeti, Cameron McCormick, Mohamed Adam, Drew Bednasek, Julie Morris, Noel Rasor, Danielle Monty-Mara, Edmond Mara, Ryan Lash, Mark Carper, and Peter Sam have at least listened to me rant about one thing or another in the process. Neither the rants nor the book would have happened without Ang's coffee, Chinyanja class, or music research fieldtrips. I have been blessed with colleagues and friends like Paul Hotvedt, Jake Gordon, Peter Ukpokodu, Suresh Bhana, John Janzen, Khalid El-Hassan, Craig Pearman, Liz MacGonagle, Steve Egbert, Terry Slocum, Kevin Price, Johan Feddema, Brian Daldorph, Peter Herlihy, Chris Brown, Curt Sorenson, Les Dienes, Pete and Barbara Shortridge, Shannon O'Lear, Gitti Salami, Naima Omar, and Byron and Marta Caminero-Santangelo with whom I have been able to share a variety of ideas. Liz, Steve, Terry, Kevin, Johan, Brian, and Byron shared some of this journey on their visits to Zambia with the State Department linkage grant. Byron deserves special thanks for his critical engagement with the interests we share, as well as for his touch of lightness and humor – always welcome on very long bus trips in Zambia, kids' play dates, or traffic detours leaving Kansas basketball games. My family, Phebe, Atlee, and most of all Melanie, have once again put up with my long absences and have given me the time and space to do what I love. I thank all three of you, especially for sharing much of the fieldwork time with me, which was a great delight. Collecting trash and picking around city dumps may not be your idea of a good time, I know, but it is hard to top some of the moments in our lives that came along with all that, in Tanzania and Zambia with the three of you.

List of Abbreviations

ASP	Afro-Shirazi Party, ruling party of Zanzibar, 1964–1977
BSAC	British South Africa Company
CBD	Central business district
CBE	Community-based enterprise
CBO	Community-based organization
CCM	Chama cha Mapinduzi (Party of the Revolution), Tanzania's ruling party, 1977-
CIP	Community Infrastructure Program in Dar es Salaam
COLE	Zanzibar's Commission for Lands and Environment of 1989–2001
CUF	Civic United Front, the main Zanzibari opposition party, 1992–
DANIDA	Danish International Development Agency
DED	German Development Service
DFID	Department for International Development in the government of the United Kingdom
ECZ	Environmental Council of Zambia
EPM	Environmental Planning and Management, the model framework for planning under the Sustainable Cities Program
FDD	Forum for Democracy and Development, a Zambian opposition party
FDI	Foreign direct investment
FINNIDA	Finnish International Development Agency
HIPC	Heavily Indebted Poor Countries initiative of the World Bank
IFI	International Financial Institutions, shorthand for the World Bank and the International Monetary Fund
IMF	International Monetary Fund
JICA	Japan International Cooperative Agency
LCC	Lusaka City Council
MMD	Movement for Multiparty Democracy, Zambian ruling party, 1991–
NAZ	National Archives of Zambia
NCCR	National Convention for Reconstruction and Reform (Mageuzi), Tanzanian opposition party since 1992
NGO	Non-governmental organization
OECD	Organization for Economic Cooperation and Development
PAYE	Pay As You Earn program in Lusaka
PIG	The Party and Its Government, Zambia's single-party regime led by the United National Independence Party

PROSPECT	Program of Support for Poverty Elimination and Community Transformation in Lusaka
PRSP	Poverty Reduction Strategy Paper
PUSH	Peri-Urban Self-Help program in Lusaka
RDC	Residents Development Committees of Lusaka
SADC	Southern African Development Community
SAP	Structural Adjustment Program
SCP	Sustainable Cities Program
SDP	Sustainable Dar es Salaam Program
SIDA	Swedish International Development Agency
SLP	Sustainable Lusaka Program
STCDA	Stone Town Conservation and Development Authority in Zanzibar, 1989-
SUDP	Strategic Urban Development Plan in Dar es Salaam
TANU	Tanganyika African National Union, mainland Tanzania's ruling party, 1961–1977
UASU	Urban Authorities Support Unit in Tanzania
UMA	Urban Management Advisors of the German Development Service in Zanzibar
UN	United Nations
UNCED	United Nations Conference on Environment and Development in Rio de Janeiro in 1992
UNCHS	United Nations Center for Human Settlements, now known as UN-Habitat
UNDP	United Nations Development Program
UNEP	United Nations Environment Program
UNIP	United National Independence Party, Zambian ruling party, 1964–1991
UPND	United Party for National Development, a Zambian opposition party
WSSD	World Summit on Sustainable Development in Johannesburg in 2002
ZCCM	Zambia Consolidated Copper Mines
ZMC	Zanzibar Municipal Council
ZNP	Zanzibar National Party of 1954–1964
ZPPP	Zanzibar and Pemba People's Party of the early 1960s in Zanzibar
ZSP	Zanzibar Sustainable Program

Glossary of Foreign Words
(Kiswahili words, unless otherwise noted)

Bongo	Brains, the colloquial nickname for the city of Dar es Salaam
Bongo flava	Dar es Salaam hip-hop style
Fitina	discord
Kapenta	small dried fish (Chinyanja)
Kapitao	captain, neighborhood headman (Chinyanja)
Kujitegemea	self-reliance
(Ma)diwani	city councilor(s) in Zanzibar
(Ma)jambazi	armed thug(s)
(Ma)sheha	local ward-level government official(s) in Zanzibar
Mwafaka	pact or agreement
Ng'ambo	The other side of a city, the other bank of a river, or overseas
Ng'ombe	cow (Kiswahili and Chinyanja)
Tarab	Swahili coast popular music style
Ubinafsi	private-ness, selfishness
Ubinafsishaji	privatization
Ujamaa	family-ness, the Tanzanian ruling party's governing ideology from 1967–1985
Ujima	collective responsibility
Umoja	social unity
Utajirisho	enrichment, the policy slogan of the opposition Civic United Front in Zanzibar
Walalahoi	the dispossessed ones
Wamachinga	street peddlers
Wanyonge	the abject poor
Wazungu	whites

Chapter 1

Toward a Political Ecology of African Cities

Look, does water have an ideology? Does a drainage system have an ideology? Does garbage?

(Mnoga 2003)

Introduction

In August 2003, I was collecting garbage in Zanzibar. I was participating in the first few weeks at work for a community-based solid waste management organization formed out of the Mkele Ward Development Committee in the city. The group's employees and I pushed wheelbarrows around the dusty rutted alleyways of Mkele ward in pairs all morning, dumping our cart-loads into the flatbed truck that the Ward Development Committee had secured a few months before this from the German embassy (see Figure 1.1). My wheelbarrow partner, Kasu Bachu, would jauntily sing out, to a tune of his own design, "bring us your garbage, we want your garbage, garbage is wealth," as we worked. Three workers were also busy reducing the overwhelming backlog of waste that had built up at the group's lone slab site, every once in a while coming across a rat or a snake in the muck. Whenever we filled the truck, four of us piled into it and took the waste to the partially controlled dumpsite that serves as Zanzibar city's landfill. On our last such trip of the workday, we would stop by the bustling roadside stalls in the peri-urban community around the dump and buy some fruits. While the donor funding of the project lasted, the employees like Kasu would enjoy the modest, steady wages – about 50 dollars a month – and they were clearly going to enjoy the fruits while they could.

One day, when Kasu and I came to his own house to collect, he insisted that I come inside and meet his family. His neighbors gently teased him about his new rich *mzungu* (white) friend as we went in the door. We sat in the small sitting room that Kasu, his wife and his young baby shared with his parents and siblings. His mother brought us some juice, made cold by the small refrigerator that Kasu's new job had helped the family to purchase. The house had few furnishings, bare concrete block walls, and a leaky rusted roof of corrugated iron. Its walls stood less than a meter from those of the neighbors on two sides, with about two meters open at the front. At the back, the plot backed up against a drainage ditch that threads its way through Mkele down a

1

Figure 1.1   Mkele's truck at the dump, Zanzibar (2003)
Source: author

slight slope toward the ocean barely a kilometer away. Water flows in the ditch all year long, and it frequently overflows when the garbage that gets thrown in it causes the stream to back up.

There is so much trash everywhere in Mkele – in the drains, in the sands, in the trees, much of it organic food wastes, but nevertheless mixed with a little gray water, battery acid, and pesticide – that even an intensive effort would barely touch the surface. Over the past few decades, Zanzibar has managed to collect and dispose of approximately 40 per cent of the solid waste its residents produce – a comparably respectable percentage among African cities. But Zanzibar, like many cities across the continent, has grown dramatically in the last forty years, from under 50,000 people in 1958 to nearly 400,000 people in the metropolitan area by 2005. This growth has come with only a meager expansion in the economic opportunities available to city residents and a chronic crisis in the provision of services like solid waste. The severe urban poverty and poor environmental health conditions in Zanzibar come attached to a contentious society that is still "searching for more democracy," as the Tanzanian urban planner Fred Lerise (2000, p. 89) deftly phrases it. Having endured Omani domination (1696–1890), British colonialism's polite repression (1890–1963), and a single-party police state (1964–1992), Zanzibar city residents have watched the first dozen years or so of a multi-party political system (1992–2005) heave to and fro in the rough seas of racial, regional, religious, gender, and class tensions.

The challenges that cities like Zanzibar and the people in them have faced have brought on wave upon wave of development agendas designed to solve them, for more than four decades now, to little effect. The latest of these waves, to which phrases like "sustainable development" and "good governance" are inexorably attached, involves a heavy reliance on non-governmental organizations (NGOs) and private sector firms in what are euphemistically referred to as "partnerships" with government agencies in the management of everyday life. Good governance partnerships for sustainable development are supposed to provide a path out of poverty toward a better environment and a society with "more democracy." For instance, the Mkele waste collectors work under the Mkele Ward Development Committee NGO, which works for profit in collaboration with the now democratically elected Zanzibar Municipal Council, under the Minister for Good Governance of the Zanzibar Revolutionary Government (itself a semi-autonomous "partner" within the United Republic of Tanzania). The money driving the project – quite literally so, in the form of the donated truck and wheelbarrows – has come from the German Development Service and the German Embassy. The Germans funded the creation of the drainage ditch behind Kasu's house, too. We look at the Mkele case more closely in chapter four. For now, think of it as spelling out some basic questions about how far the new vocabulary of development has taken the urban poor majority along the path to a better life. Frankly, the first surface of an answer seems to me to be fairly simple: not very far. Garbage is only wealth for a very select few. Others, like Kasu, have decent jobs for a while, and little reminders of what could be, like the small refrigerator that made our juice cold. A few more tons of waste get collected for a while, in one of the more than fifty poor and under-serviced wards of the city. But no one involved in the Mkele project in 2003 was under any illusion that anything more would occur beyond a temporary dent in the fender of the steamroller of urban problems, to say nothing of global and national forces, that Mkele and Zanzibar face. The second answer, then, involves explaining a context with so few illusions or delusions of development left.

In this book, I want to use questions about solid waste to explore what it has been like to manage, to cope, and to live in cities like Zanzibar, in wards like Mkele, for people like Kasu and his family, in the last few decades. My main emphasis is on an examination of outcomes from the United Nations Sustainable Cities Program (SCP), the program responsible for the creation of the project in Mkele that Kasu works with. I examine three case study cities – Dar es Salaam and Lusaka as well as Zanzibar – in later chapters, emphasizing their SCP experiences. In this chapter, I lay out the broader context of the urbanization dynamics of Sub-Saharan Africa, and four themes that define African development over the past fifteen years or so, which I label as *neoliberalism, sustainable development, good governance*, and *the politics of cultural difference*. I end the chapter by discussing the approach that I take in the book to understanding how these four themes intertwine in the cities.

African Cities

Sub-Saharan Africa is, like Zanzibar, becoming urban. The world's least urbanized countries have been the most rapidly urbanizing ones for the decades since the 1960s. Eastern and southern African countries have led the world in urbanization rates for nearly half a century. Even if we accept that aggregate national statistics from Africa are often somewhat problematic, it is undeniable that the urban populations of Sub-Saharan Africa, and eastern and southern Africa especially, have grown at a substantial pace. A dozen eastern and southern African states had annual rates of growth for the urban population that were more than twice the annual rates of natural increase for the overall population of the country, both for 1960–91 and 1991–2000 (Simon 1997, p. 88). Tanzania, Botswana, and Swaziland had annual rates of urban population growth near or above 10 per cent from 1960–91, with Mozambique at nine and a half per cent a year. Even during the 1990s, when many observers noted a slowdown in African urbanization, several countries had estimated urban growth rates near or above seven per cent. In less than 50 years, some eastern and southern African countries have gone from being largely rural societies to being places where almost half the people live in or around cities (see Table 1.1).

Despite claims that are often raised that development assistance has displayed an "urban bias" in Africa, meaning a bias against rural areas and in favor of cities in aid moneys, it is still apparent that Sub-Saharan African urbanization processes and urban dynamics are poorly understood and comparatively under-studied (Becker et al. 1994; Simone 2004). This enduring gap is particularly unfortunate both because of the rapid growth of many cities and because the path of urbanization in most countries has been somewhat distinct from what has been seen in other regions of the world, particularly in wealthy European or North American settings. With some exceptions, the extraordinary story of urbanization in Sub-Saharan Africa has not accompanied a substantial economic transformation of society towards industry and manufacturing (Inoguchi et al. 1999, p. 2). Certainly in some countries, notably South Africa, industrial development occurred with the urban-ward trends. But in much of the region, the ever-expanding numbers of

Table 1.1 Urbanization rates for selected eastern and southern African countries

COUNTRY NAME	Average annual percentage growth of urban population		
	1975–79	1980–90	1990–98
Botswana	8.8	14.3	5.3
Kenya	8.6	7.9	6.3
Lesotho	7.1	6.8	5.8
Malawi	7.1	7.1	9.5
Mozambique	12.1	9.2	6.9
Tanzania	**11.6**	**6.9**	**8.0**
Zambia	**6.3**	**3.1**	**2.8**

Source: World Bank, *African Development Indicators 2001*, Washington, DC: World Bank, p. 311

urban residents have as a consequence of the stagnation of manufacturing and formal industry become increasingly dependent on what are termed informal activities – small scale, low-technology manufacturing, petty wholesale trading, and informal service provision – for basic needs and daily life. Land is purchased, transferred and developed, houses are built, commodities produced and consumed, and garbage collected and disposed of outside of the purview of states and often in technically illegal ways by nearly everyone in many of these cities.

African urbanization since the 1960s also has taken place amidst a total sea change in African politics and society, and a time of economic, environmental, political, and cultural transformation in the world. It is impossible to pin down all of the swirling forces intersecting in African cities, but four different story lines come to the fore that relate to my focus in this book: neoliberalism, sustainable development, good governance, and the politics of cultural difference. I elaborate on, and show how the case study chapters are organized to approach each, in the following four sub-sections.

Neoliberalism

Having gained political independence (mostly in the 1960s) without economic autonomy, most Sub-Saharan African countries followed modernization strategies largely dictated by the Western powers, whatever rhetorical flourishes to the contrary the new leaders offered. With the steady collapse of many primary product prices (upon which many of the continent's economies still depend), the oil shocks of the 1970s, and the profligate government spending expected in various development models of the era, the majority of countries fell into severe public debt. The debt crises led to the imposition of Structural Adjustment Programs (SAPs) by the International Financial Institutions (IFIs) beginning in 1981. Structural adjustment itself adjusted its terminology in the 1990s, as the IFIs advocated – and required – Poverty Reduction Strategy Papers (PRSPs) to orchestrate modest debt relief for the Heavily Indebted Poor Countries (HIPCs) that they held in their sway.

Most Africans at the beginning of the 21st century thus find themselves continuing to live in a time of what geographers Richard Peet and Michael Watts (1993) termed "market triumphalism." Market triumphalism involves a globalization of a SAPped, PRSPed, and HIPCed ethos, in which "individual and corporate freedom, and market-oriented production and consumption, have been considered sacrosanct" (Inoguchi et al. 1999, p. 3). The now-ubiquitous development model termed neoliberalism combines a heavy emphasis on exports contributing to an overall outward orientation to the economy, a rollback of government intervention in economic matters, a devaluation of national currencies, a welcoming stance toward foreign capital, and a devotion to free markets. The apparent sameness in African development planning policies results from pressures from the IFIs and western countries, the narrow range of policy options available to governments which sign adjustment agreements, and the shortage of clearly articulated alternatives to capitalism (Simon et al. 1995; Minogue 2002).

Sub-Saharan Africa is often figured to have experienced exceedingly limited impacts of economic globalization because of the paltry percentages of global foreign direct investment (FDI) flows to it (Briggs and Yeboah 2001). It sometimes appears as if "the global development cartography has been redrawn, and the new map excludes Africa" (Edoho 1997, p. 12). Yet due to the neoliberal IFI policies, Africans experience globalizing forces in political-economic and socio-cultural terms whose impacts may be proportionally greater in some ways. Global forces are attempting quite literally to reconstruct the entire political economy of Africa in their image. Some pessimists go so far as to claim that "the neoliberal stranglehold in Africa ... has reduced or eliminated the integrity of what were fragile nation-states and pushed them in to anarchy and civil wars" (Hoogvelt 1997, p. 163).

Even where anarchy and civil wars have not transpired, neoliberalism has had substantial implications for African cities. Neoliberalism is clearly a large part of what led the exasperated former mayor of Zanzibar to ask me the rhetorical questions with which I began the chapter: it seems that water, drainage, and garbage do, indeed, have an ideology attached to them now. In Tanzania, the convoluted Kiswahili term for this ideology is *ubinafsishaji*; it is translated as "privatization," but might be creatively reworked in English as "the causing of individualism or selfishness." The privatization of the economy and privatization of service delivery have changed how people work and how they get to work in cities across the continent, as well as how they take care of their garbage (Bartone 2001; Wolgin 1999). The high rates of urban unemployment produced by rollbacks on government jobs – and not substantially reduced by any parallel growth of private sector employment – has led to a spreading informality to work and to many aspects of life (Hansen and Vaa 2004). Informality increasingly dominates the delivery of urban services such as water supply, solid waste management, and infrastructure provision as public and private sector efforts falter in marginal majority areas of the cities. City governments are often severely hit by neoliberal policies, particularly in regard to fiscal matters (Kironde 2001a). Although structural adjustment reformers often advocate a decentralization of politics, the same free marketers strip away, as "nuisance taxes," the very revenue sources that give city governments the limited degree of autonomy they've earned. The privatization of service delivery – often a linchpin of the UN Sustainable Cities Program – does remove some of the financial burdens from urban authorities, but it also takes away basic revenue sources. Central governments – the main sources of revenue for urban governments – operate by reducing or withholding budget payments from cities, further draining them of any capacity.

In the larger scheme of things, what seems to be happening is something startlingly close to what Watts (1994) once called "the privatization of everything." Neoliberalism's triumphal ethos has wormed its way well beyond economic and governmental policy. The privatization of culture, society, and consciousness has reshuffled ethnic and family relationships toward a more individuated, and fractured, sense of self and a commodification of many aspects of everyday life (Rapley 2004, p. 64; Kelsall 2004, p. 70). In the end, every actor in the city of *ubinafsishaji* has cause to be selfish, if only just to get by.

In each case study chapter, I examine the particulars of neoliberalism's imposition in that city. I detail the events and processes behind its inter-relationship with the UN Sustainable Cities Program, and particularly through the lens of solid waste management policy.

Sustainable Development

Alongside neoliberalism, contemporary African cities have also played host to another powerful global development agenda, in the form of sustainable development. Sustainable development is a phrase that has taken on a life of its own, or, really, several lives (McGranahan and Satterthwaite 2002; Bucking-ham-Hatfield and Percy 1999). The phrase has come to have many meanings in global documents, all of which vaguely hover around a balancing of economic growth and environmental impacts from one direction or another. Most globally influential texts of environmentally sustainable development are "vague on operational issues" and highly contradictory (Barrow 1995, p. 371). Local socio-political relations and the bureaucratic cultures implementing urban policies have enormous importance to policy outcomes in addressing this contradictory idea of sustainability. Yet in mainstream literature on environmentally sustainable development "the political dimension typically is ignored" (Bryant 1991, p. 164).

There may be a reason why the political dimension is ignored: in most cases, as Catriona Sandilands (1996, p. 124) says, "sustainability is considerably more politically palatable precisely because it obscures important problems of definition, vested interest, and ideological conflict." Economists have sought blandly to model it as "constancy of the natural capital stock," meaning "non-negative change in the stock of natural resources and environmental quality" (Pearce and Turner 1990, p. 4), but this attempt at simplicity is deceptively complicated. For instance, who gets to define what a "non-negative change" is? Sustainable development can throw around a lot of ideological and political might depending on how it is defined, or which element of the definition takes greater precedence. Keeping knowledge of the exact meaning of the phrase obscure, in fact, may serve the interests of those in power.

Sustainable development language has come to be applied more and more frequently to urban settings, in both the developed and developing worlds' cities (Girardet 1999; Simon 1999a). For developed world settings, sustainable development seems to mean finding a way to cope with industrialization and urbanization's "twin perils: social and economic destitution and massive environmental degradation" (Hanson and Lake 2000, p. 2). But for the dominant world-view of developed world elites, sustainable development means, as Susan Hanson and Robert Lake (2000, p. 2) put it, "embracing the agenda of the market, top-down planning, and scientific technological, and/or design-based solutions" to environmental problems. Sustainable development often becomes a new neoliberalism. Then US Secretary of State Colin Powell (2002, pp. 6–7) made this agenda clear in addressing the National Academy of

Sciences annual meeting in 2002, on his way to the World Summit on Sustainable Development in Johannesburg, South Africa. To Powell (2002, p. 7), sustainable development meant combining "a new approach to global development, designed to unleash the entrepreneurial power of the poor" with "good governance, sound institutions, economic reform, transparency in your system, the end of corruption, responsible leadership, responsible political activity, and ... decision-making based on sound science."

By contrast with the likes of Secretary Powell, to most African studies urban scholars, as well as many urbanists in the developed world, sustainable development emphasizes "sustainable livelihoods at the local-scale" (Hanson and Lake 2000, p. 3; see also Lerise 2000; Rakodi 2002b). Like many Africanists, I prefer a livelihood approach to sustainable development. But I contend in this book that if we want to understand the operationalization of the other, more dominant vision of sustainable development in African cities – as in the example of the UN Sustainable Cities Program – we can best do so through careful attention to the question of "governance." Hence in the sustainable development sections of the case study chapters, I look at the environmental governance policy networks of the cities alongside the environmental fallout of them for the livelihood of ordinary city residents.

Good Governance

In the same decades that have seen the rise of neoliberal development policies and sustainable development rhetoric, Africa has experienced broad move-ments for political change. The most visible manifestation of this climate of transition is the re-emergence of multi-party political systems. Tanzania's single-party system that had operated formally since 1967 ended officially in 1992; similarly, Zambia's experiment with a single-party system that had formally begun in 1972 came to a close in 1991. Less visible, but perhaps more significant strands of political openness are evident in countries like Tanzania and Zambia, with the emergence of a broader popular press and non-governmental television and radio outlets. A loosening of some aspects of state control on associational life has resulted in a proliferation of non-governmental development organizations, or what is characterized as civil society. Donor countries and the IFIs have championed civil society, their idea of multi-party western liberal democratization, and the empowerment of community-based organizations like that of my friends in Mkele, in the process often attempting to circumvent states and governments. A more politicized and more vocal populace in marginal states often weakens African governments as much as it strengthens them, leading many governments to see the organized public in enemy terms. This phenomenon is most in evidence in urban areas. Multiparty democratic transitions have often been triggered by popular action in urban areas, and urban areas regularly vote to spearhead opposition movements. There are numerous cases across the continent of central governments whose legitimacy is undermined by significant unpopularity in major cities, and even a few cases of city councils controlled by different parties from the central governments. In the context of central government control over the fiscal allotments to cities, a scenario that involves the further erosion of an opposition city council's authorities and capacities becomes all the more likely to transpire.

The weakening of state authority and strengthening of civil society non-governmental agency while market forces are let loose to do as they will can create the very definition of what neoliberals see as "good governance" (Batley 2001; Griffin 2001). To many neoliberals, good governance just seems to mean "less state intervention" (Drakakis-Smith 1997, p. 815). Good governance is a fairly slippery concept, though. Hyden and Mugabe (1999, p. 33) note that "the call for good governance ... usually implies a whole range of issues," and not necessarily a neat confluence of capitalism, bourgeois democracy, and efficient technology. Within this rubric of good governance, in many countries, particularly in urban settings, new forms of "partnerships" have come to the fore (Mercer 2003).

As the case studies in chapters three to five suggest, non-governmental organizations and the ordinary citizenry sometimes do find their voice in urban management agendas of good governance, through "stakeholding" and other innovations (Inoguchi et al. 1999, p. 8). Yet good governance often leads "not to democratization but to the disempowerment of local authorities and local societies alike" (Beall et al. 2002, p. 65). For this reason it can ultimately be understood as a rhetorical device that reproduces an *exclusionary democracy* in African cities, to borrow the phrase from the political scientist Rita Abrahamsen in her book, *Disciplining Democracy*. Abrahamsen (2000, p. 50) points to the contradictions in the way the World Bank discusses governance in Africa. In Abrahamsen's terms, exclusionary democracy is a deliberate product of these contradictions. Exclusionary democracies "allow for political competition but cannot incorporate or respond to the demands of the majority in any meaningful way" (Abrahamsen 2000, p. xiv). Abrahamsen contends that the World Bank has utilized a one-size-fits-all agenda for good governance as a means by which to further secure control over African states in the interests of the penetration of global capital and extraction of industrial resources. This is becoming a fairly common claim in African studies, as part of what Dani Nabudere (2000, p. 23) calls the "third colonial occupation." Zambia, for example, is said to be "more foreign-owned" now than it was at independence (Tandon 1999, p. 35). Abrahamsen's "good governance agenda" allows the World Bank to appear as if it is speaking for the ordinary people and their empowerment, threading its unpopular policies in with the popular movements for political pluralism across the continent.

Abrahamsen's critique is sweeping and convincing, and it has influenced my take on governance in this book. But there are some gaps, in fairness somewhat due to the design of *Disciplining Democracy*. For my purposes, the problems are that Abrahamsen does not consider urban areas, the rhetoric of other donors beside the World Bank, or the colonial precedents of the good governance rhetoric she critiques. I tackle the first problem very simply in this book, by taking her arguments and inspiration into the three cities. In chapter two, I address the second issue by looking at the rhetoric and discourse of the northern European donors who dominate the funding base for the activities of

the Sustainable Cities Program. Abrahamsen (2003, p. 195) herself acknowl-
edged the third issue in more recent work, stating quite plainly that the
"colonial experience is ... crucial to an understanding of contemporary
politics." This point is the central theme in the work of another prominent
political scientist, Mahmood Mamdani. In his book, *Citizen and Subject*,
Mamdani (1996, p. 6) highlights the struggle in late colonial times that resulted
from the attempts of European colonial states to establish what he terms
"decentralized despotism" as a form of institutional segregation. Like
Abrahamsen, Mamdani makes the idea of exclusionary governance central
to his arguments, developing a sophisticated picture of the "bifurcated state"
under colonialism. There was, in a nutshell, a state for citizens (meaning whites
and elites) and a state for subjects (meaning everybody else).

Recognizing, with Pal Ahluwalia (2001), that there are dangers in making
too much of the direct legacies of colonialism, it still seems clear that one of the
most critical legacies of late colonialism lies in the tactics and strategies of the
colonial states in cities as the "subjects" came to reside in them in droves. The
bifurcated state of late colonialism nearly blew apart because of its struggle
with the contradiction of citizens and subjects residing together in cities. In
postcolonial cities, the categories are not as sharply distinguished, but state
powers certainly continue to exercise very similar exclusionary tactics,
repeatedly. African postcolonial states often resort to repression and violence
rather than policies to achieve legitimacy, and their legitimation is of a sort that
Achille Mbembe (2001) terms "private" and "indirect": in other words,
ubinafsishaji. Even in the "democratization" of the past fifteen years,
alternative or progressive voices and agencies "have often been ignored,
bypassed or undermined" (Simon 1995a, p. 36). As the case study chapters
show, many postcolonial states in Africa have become "private" arenas for
individuals within them to pursue their own accumulation strategies, as in solid
waste enterprises begun within the rubric of the Sustainable Cities Program.
The clientele who benefit from their largesse give the regimes they serve their
strongest hold on legitimacy in the context of dubious electoral politics
(Cammack 2002). States operate "indirectly" because structural adjustment
policies and neoliberalism have taken them out of direct spheres of control.
The new wave of democratization throughout much of Africa in the 1990s,
coupled with the broader idea of a good governance agenda, thus becomes
much more clearly visible in the light of the exclusionary geographies that
pattern the landscapes of its cities, as I aim to show in the case study chapters.

The Politics of Cultural Difference

The inflamed passions of racial, religious, gender, and class conflict have
accelerated dramatically in much of Sub-Saharan Africa over the past few
decades. Many of the conflicts originate in political struggles over socially
constructed and contested ideas of difference between and within cultures. The
result is often a more literal – and bloody – example of the "culture wars" that
have created rifts within American society during this same time period
(Mitchell 2000). Some of the most horrific or infamous examples have been

generally rural – such as the Rwanda genocide or the Congo civil war. Much of the rest of the passions have been urban.

Not all of the sweeping cultural changes of the past two decades have been first and foremost conflictual. I am thinking, for instance, of the "massive growth of revivalist churches [mosques] and sects, whose gathering places provides redemptive moments and spaces of spiritual and social communion" (Zeleza 1999, p. 55). Yet even these religious havens – indeed, sometimes especially these spaces – are central to small local conflicts. Little skirmishes in the African versions of the culture wars taking place around the globe happen every day in the continent's cities. To wit, in January 2003, a melee erupted on Chachacha Road in the Lusaka Central Business District after a group of Christian extremists started forcibly stripping the clothing off of downtown businesswomen they deemed to be so inappropriately dressed as to be taken for prostitutes. In August 2003, rioting Muslim extremists in Dar es Salaam occupied a Muslim burial site that they claimed the city government had desecrated by planning an office building that required the cemetery's relocation. In March 2004, two bombs reputedly placed by the members of a religious NGO exploded near the homes of prominent Zanzibari government officials; a third bomb was defused outside a tourist restaurant. In each case – and there are hundreds of cases like these across Africa lately – it would be easy to read the culture-war issue as one of religious fundamentalists standing against the surging tide of a global secular order. In other words, what we are looking at is a "clash of civilizations," a conflict of "tradition against modernity," or "the cross versus the crescent" (Mbogoni 2004; Schech and Haggis 2000). These sorts of conclusions often come readily from analyses of Islamism, but just the Lusaka example alone that I've given demonstrates that it would be a serious mistake to wrap up all of the isms of the restless billions into Islam (Watts 2003). The clashes of Dar es Salaam have been as much the work of followers of the "cross" as followers of the "crescent." These cases, and many like them in Sub-Saharan Africa, present deeper and more complex stories once they are examined a little more carefully.

Each of these examples is built on a history of colonial race- and class-based segregation, highly fractured gender and identity politics, and regional discrimination that in many ways intensified after colonialism's demise. Zanzibar, Dar es Salaam, and Lusaka all experienced the racial segregation that served as the foundation of British colonialism in urban areas, and the physical and mental dimensions of these segregationist tactics have proved quite difficult to just wipe away. Colonialism worked on the basis of the divide-and-rule, exclusionary tactics Abrahamsen and Mamdani have highlighted; it proved all too simple for postcolonial regimes to replicate and expand on them, to use the invented traditions and identity politics of colonial times to their advantage (Schech and Haggis 2000, p. 143). Lusaka is still very much bifurcated between "yard people" (those who live in the rich, formerly white areas) and "compound people" (who live in the generally poor, formerly "unauthorized compounds" of the colonial map; see Figure 5.2). Dar es Salaam is still generally a city with the colonial town plan's zones A (white elite), B (Asian middle class), and C (African proletariat) on the brain and on the landscape. Zanzibar has its tourist haven of Stone Town and beach villa zones north and south of the city on one regime of power grids, sewerage, schooling, and privilege, and most of its Ng'ambo (literally, its "Other Side") on quite another.

Without condoning the violence in these examples above, it should be noted that each involves an act or acts committed by an ostracized and poor minority element of city life – albeit often orchestrated by a well-educated clique within that element – coming out of slums and squatter areas (the compounds, Area C, or Ng'ambo), against established elite institutions. Each can be construed to have ties to the perceived threats to these almost exclusively male elements by the liberation or empowerment of women that they see as being promulgated by that establishment. Each has the possibility of connecting with long histories of regional division. The radicals in Zanzibar are portrayed as sympathetic to Pemba-islander culture. Many extremist Muslims in Dar es Salaam are drawn not from traditionally Muslim coastal ethnicities but from recent converts in peri-urban informal settlements, often from poor, rural regions upcountry. The moralizers in Lusaka came mainly from local culture groups threatened by the huge wave of in-migration to Lusaka by the more highly educated, upscale cosmopolitans of the Copperbelt in northern Zambia or the new monied white South African investors. This very complex politics of cultural difference – where cultural difference is a political weapon and shield – is manufactured and then manipulated by various players with varying degrees of success, some from each of these angles.

This politics of cultural difference in African cities can be seen as stemming partially from colonialism. Colonialism took "a multitude of African social formations with different, often particularist memories competing with each other" (Mudimbe 1994, p. 129) and sought to meld these into a coherent map of subordinate groups. It *domesticated* differences, meaning it made them appear as common sense everyday reality. It made them manifest domestically, in the types of houses and locations of those types, and in the broader ethnic, racial, regional, religious, class or gender distinctions that furthered its interests. Colonialism certainly doesn't explain every point of origin for the politics of cultural difference today. But it does offer the vital reminder to any inquiry into Sub-Saharan African cultural geography to situate African cities today between the warp of colonialism and the woof of the post-colonial inheritors of its powers.

The fact that the development projects of post-colonial inheritor states so frequently failed to bring broad improvements in the quality of life – coupled with the inequalities exacerbated by neoliberalism and the exclusionary democracies fostered by the good governance machinery – fueled the explosive potential of the Other Sides of African cities (Watts 2003). If we think of culture as something practiced and not owned, then we can say that its practice has, in African cities as in many places, become even more substantially contested than in previous epochs (Shurmer-Smith 2002). These cities are experiencing "battles over cultural identities – and the power to shape, determine, and literally *emplace* those identities" (Mitchell 2000, p. 11). Hence in the city chapter sections dealing with the politics of cultural difference, I make the case that understanding those battles requires us to understand their historical, political-economic and environmental contexts.

Political Ecology and Urban Africa

The ongoing politics of cultural difference, then, demands of my study in this book not only a lively cultural-historical imagination, but one that pays attention to politics, economics, and the environment at the same time. In a phrase, to me that means doing what these days gets called political ecology. Political ecology grew out of the frustrations of human-cultural geographers and anthropologists with the limitations of purely physical analyses of environmental degradation. Political ecologists have sought to place degradation processes in their "historical, political and economic context" (Blaikie and Brookfield 1987, p. 17) as well as the ecological, in effect to merge cultural geography or cultural ecology with political economy in order to study the environment. Connecting these two approaches is so necessary precisely because all of urban Africa's environmental problems, or what might more properly be called environment-and-development problems, take place in what Raymond Bryant and Sinead Bailey (1997, p. 39) have described as "a politicized environment." The chief contributions of this recognition are that power relationships are inherently unequal in the dynamics of environmental change, and that this systemic inequality of power relates to systemic environmental outcomes (Hardoy et al. 1992). But there are several aspects of a political ecological approach that need further refinement. One problem has to do with the scope of analysis. Political ecology's practitioners are sometimes accused of a romance of the local, when environmental issues operate "simultaneously at the local, regional, and global scales" (Bryant and Bailey 1997, p. 194). Scales are socially constructed and politically reproduced, and not innocuous containers (Zimmerer and Bassett 2003, p. 3). I try to address this by not only assessing three different case studies but by seeing their links with each other and the nested (global-national-local) and contested scales of urban politics and policies.

The second issue has to do with urbanism. Although a fair number of studies – many of them produced in Africa by African academics – have addressed urban environmental issues from critical political perspectives, works explicitly articulated as political ecology in Africa have – with a few exceptions (Friedberg 2001) – overwhelmingly maintained a rural focus. The preponderance of rural studies in Africa does make some sense given the dominant location of the African people, in rural areas. But the ironic result is that a great deal of the political ecology research that does take place in urban settings has a Euro-American focus, or, more recently, a Latin American one (Swyngedouw 2003 and 2004; Pelling 2003a and 2003b). This seems unfortunate to me, when African cities are experiencing such new forms and still-rapid rates of urbanization and critical studies of them end up under-examined in the broader global literature of urban studies (such as Kironde and Ngware 2000 or Aina et al. 1994). Our understanding of African urban environmental issues can benefit from the integrative, multi-dimensional language of political ecology research, attending to "unequal power relations in the context of the urban environment" as I seek to in the book (Bryant and Bailey 1997, p. 194).

Thirdly, the dual pulls of global and local scales can also mean that the national/state dynamics are shortchanged in political ecology. As Richard Schroeder (1999) shows, both the global and the local scale construction of environmental conservation in Africa depend upon national states for implementation at a time when authority at that mediating scale is being both strengthened and weakened. There is a need to be much more attentive to the complex politics in which African states and their agents engage (Watts 1987, 1993, 1997, and 2003; Mohan and Stokke 2000).

To put the reasons for this need more plainly, let us consider the question of solid waste. Solid waste can be seen as a "livelihood resource or a health hazard," or it can simply be ignored along the sides of roads and alleys as piles of it build up or burn down (Pelling 2003a, p. 74). How it is perceived will vary in any African city, but the "persons and institutions that control the content of political discourse" will shape the policies and relations designed and enacted to deal with it (Pelling 2003a, p. 74). And that means the bureaucratic culture of a city, its distinctive and ever-changing character of government-society relationships, will have much to say about problem perceptions and policy priorities regarding garbage (Gandy 2002, p. 5). Some combination of central and local state agencies is crucial to the bureaucratic culture even in countries where privatization has proceeded at breakneck speed. I therefore attend to the details of government-society relations, particularly for local government, in each of the three cities.

Cultural analysis in urban political ecology requires a delicate balance of local, global, and intermediary levels. Like Donald Moore (1997, p. 103), I contend that we must combine rigorous detail of political, material, symbolic, and historic livelihood struggles for people in African cities with consciousness of the "presence of state functionaries" and connections to distant "mines and cities" by which those cities are also constructed. This is the robust form of political ecology I aim for in the book.

Conclusion

The extraordinary story of urbanization in eastern and southern Africa over the last fifty years has not received quite nearly the scholarly attention it deserves. Part of the reason why greater analysis is needed has to do with the remarkable confluence in African cities of the grand agendas for development emanating from the West that are economic, environmental, and political in orientation: neoliberalism, sustainable development, and good governance. Another reason, though, is the re-articulation of these agendas by urban Africans who engage them at a time of astounding change, in contexts highly charged by a complex politics of cultural difference.

I began this chapter with a vignette about Mkele, a tiny ward of Zanzibar city, in the process suggesting that it is possible to see all four of these story lines intersecting as a small non-governmental organization takes to the alleys to collect the garbage. In this book, I use case studies like that of Mkele and Zanzibar to illuminate how these themes interconnect in solid waste management, and how central an understanding of them all together is to any strategizing on how to confront them. I organize the case study chapters around sections that specifically deal with each of the four story lines in each city as a means of distinguishing analyses of the economic, environmental, political, and cultural dimensions, recognizing that each arena connects directly with the others. I therefore develop my case studies as explicit examples of political ecology because these cities represent highly politicized environments, where global economic structures, unequal power relationships, and fractious cultures are embedded in the dynamics of environmental problems associated with solid waste.

I am keen to expand on political ecology's engagement with urban areas and with the fluidity of state dynamics. The processes of state formation and reconstitution associated with and occurring around the Sustainable Cities Program in the cities I examine are often taken to be creating new geographies of governance. To the extent that this is so, though, I contend that the new forms are rarely as progressive or liberatory as the rhetoric surrounding them. Neoliberal privatization often seems to sow discord and selfishness; sustainable development programs seldom improve either the environment or the livelihood of the poor; good governance recreates and improves upon the exclusionary democracies of late colonialism; and the politics of cultural difference produce debilitating battles over emplacing identity that usually leave the Other Sides of cities right where they were.

On the surface of the fantastic rhetoric, though, the ongoing moment in urban management in Africa is one that apparently offers great opportunities for the creation of inclusive cities, where the realm of the possible is wide enough to encompass rather participatory visions of neoliberal sustainable development and good governance. Ultimately, in part because the realm of the possible right now is the realm of the coin, this ongoing moment is home to a grand contradiction: policies explicitly claimed as building blocks for inclusive cities become part and parcel of exclusionary democracies instead. Chapter two unravels this contradiction at the heart of the United Nations Sustainable Cities Program, while setting that program and its backers in a broader institutional context.

Chapter 2

The Sustainable Cities Program and African Cities

The development goal of this campaign is to contribute to the eradication of poverty through improved urban governance. The vision is to help realize the 'Inclusive City' – a place where everyone, regardless of wealth, gender, age, race or religion, is enabled to participate productively and positively in the opportunities that cities have to offer.

(Tibaijuka 2001, p. i)

Introduction

This book has quite a bit in it about garbage. But it is as much about governing and managing three African cities, and the ways ordinary urbanites cope with the plans and policies of neoliberalism, sustainable development, and good governance in a time fractured by a politics of cultural difference. To keep that framework manageable, I focus a great deal of attention on analyzing the specific workings of the United Nations Sustainable Cities Program (SCP) in three case study cities – Dar es Salaam, Zanzibar, and Lusaka – wherein solid waste management has been the SCP's priority issue.

The SCP is one of the major programs tied to the United Nations network for addressing what is termed Agenda 21 for human settlements. Agenda 21 was created at the UN Conference on Environment and Development (UNCED) in Rio de Janeiro in 1992, then refined at the so-called "City Summit" (Habitat II) in Istanbul in 1996 and the World Summit on Sustainable Development (WSSD) in Johannesburg in 2002 (National Research Council 2002). The SCP has led the way in what gets called localizing Agenda 21 in Africa, meaning adapting the global document's blueprint principles to the specific circumstances of different countries. The UN Environment Program and UN-Habitat (formerly the UN Center for Human Settlements, UNCHS) – both of which have their headquarters in Nairobi, Kenya – are jointly responsible for creating the SCP, but it has become essentially a Habitat office. In this chapter's next section, I detail this UN map of engagement with African cities and introduce the Sustainable Cities Program in Africa, with a specific eye on my reasons for choosing Tanzanian and Zambian cases to study.

My other intention in this chapter is to put the broader canvas of development assistance and development rhetoric under a microscope. What are the UN offices and the donors that support them doing, and why? Critiques of neoliberalism generally zero in on International Financial Institutions (IFIs), as is the case, for instance, with Abrahamsen's analysis of the World Bank discussed in chapter one. The aid processes for most bilateral donor countries, when they do come under scholarly scrutiny, are portrayed as speaking a common language with the IFIs during the last decade. The UN is often cast aside as a fairly toothless tiger in all of this, full of airy documents and, when push comes to shove, merely doing the work of the IFIs and major donor nations. The United States, in particular, is often said to really be "calling the shots" at the UN, to borrow the title of a recent book that makes just that claim (Bennis 2000). The UN program for localizing Agenda 21 appears to its critics to have a "credibility gap" – behind its pretty words, we find "an overarching allegiance to unsustainable international development priorities premised on the imperatives of economic growth, market liberalization, and the propagation of the pseudo-political consumerist culture" (Lubelski and Carmen 1999, p. 110).

But is there more to the UN's programs like the SCP? Critical analysis suggests that the SCP, like other Habitat programs, "gradually changed towards" an outlook not too dissimilar from the World Bank's highly neoliberal "enabling approach," but a rhetorical nod toward themes of inclusion, justice, or participation is still there (Jenkins and Smith 2002, pp. 133–35). After all, although there is some Japanese, US, French, and German money in the mix, most of the funding base for the operation of the SCP in eastern and southern African cities has come from the aid agencies of Denmark, Norway, Sweden, and Ireland. None of these countries have anything of a colonial legacy in the region, nor can we really speak of an extensive Danish, Norwegian, Swedish, or Irish role in the capitalist exploitation of the region's resources. Indeed, a very thorough book series recently explored the roles of Nordic countries in the *liberation* of southern Africa over the past forty years, not in their oppression (Eriksen 2000; Sellstrom 1999a and 1999b; Soiri and Peltola 1999). Are the UN SCP offices and these specific countries' aid agencies mere tools of the neoliberal world order, carrying out its agenda of sustainable development and good governance, or do they have a broader vitality and engagement with progressive social change? How do the UN's Africa-based offices and their programs, as well as the aid agencies of countries that have generally leaned toward third-way democratic socialism at home, intersect and engage with the IFI development agenda (neoliberalism, sustainable development, and good governance) and the politics of cultural difference all around them? In this chapter's third and fourth sections, I seek to interrogate these questions. The second section, below, outlines the SCP's work in Africa and my choice of case studies.

The SCP Map of Operations

As of May 2005, the SCP claimed some form of operation in 25 countries in the world. There are longstanding programs in India (Chennai (Madras)), Chile (Conception), and Poland (Katowice), but thirteen of the countries with participating cities are in Africa. Three of the four countries where participation is at the national level are African countries (Egypt, Nigeria and Tanzania – the Philippines is the fourth country). Fully 29 of the 45 participating cities are African ones. Outside of cities in Tanzania and Zambia, the most active African cities in the program over the past decade are Accra (Ghana), Dakar (Senegal), Ibadan (Nigeria), and Ismailia (Egypt). The southeastern quadrant of Africa is by far the greatest center of density on the world map of the UN's sustainable cities. The SCP office itself is in Nairobi, Kenya. Nakuru in Kenya, Maputo in Mozambique, Lilongwe and Blantyre in Malawi, Maseru in Lesotho, Lusaka and Kitwe in Zambia, and Cape Town in South Africa join all of Tanzania's municipalities in SCP activity. Lusaka, Lilongwe, Blantyre, and Dar es Salaam have all been 'Demonstration Cities,' the highest profile of participation, at some point in time (see Figure 2.1). That gives eastern and southern Africa four demonstration cities, out of fourteen in the world as a whole.

No other region of the world has been nearly as active in the program. It is striking, for instance, to see that much more heavily urbanized countries and regions with much greater international attention to their urban environmental problems – Latin America and Southeast Asia, especially – are weakly represented. The uneven global map of participation begs the question of why eastern and southern Africa in particular has become so active in the SCP. This density of participation might be linked in part to the expanding interchange of development information in the region that has come with the Southern African Development Community (Sidaway 1998). Perhaps the profundity and immediacy of concerns for rapid urbanization and urban environmental destruction, and the implications of these for the neoliberal and good governance agendas in eastern and southern African cities, loom larger (SARDC 1994; Devas and Rakodi 1993).

After all, we have seen in chapter one that the eastern and southern portions of the continent are identified as the most rapidly urbanizing zone on the planet for most of the last thirty years (Brunn et al. 2003; Drakakis-Smith 2000). The rate of population growth for cities in the region is estimated to have slowed in the past decade, but it is still very high. Moreover, the urbanization of land has continued to accelerate, since land and housing are among the only secure investment opportunities urbanites can operationalize within the current development dynamics (Briggs and Yeboah 2001, pp. 21–2). Most cities in eastern and southern Africa and indeed Sub-Saharan Africa as a whole are experiencing rapid urbanization of land especially at the margins and edges of cities. This rapid urbanization of marginal land, the weak capacities for controlling that process, and the ecological consequences of it are among the most pressing environmental concerns in many southern African urban areas (Swilling 1997; Kironde 1999).

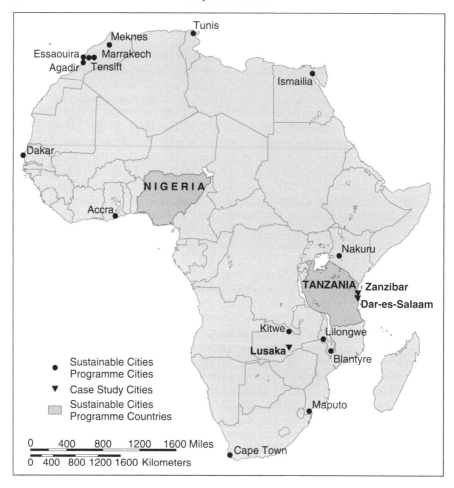

Figure 2.1   African participants in the UN Sustainable Cities Program
Source: University of Kansas Cartographic Services

Solid waste management is virtually nonexistent in these margins and edges, and scarcely shows up elsewhere in many cities of the region. It is not surprising then that solid waste has been the top priority issue identified in eight of the most active SCP cities in Africa, including the three case study cities for this book, as well as Accra, Dakar, Ibadan, Ismailia, and Moshi, Tanzania. In many participant cities, solid waste management systems in existence before the SCP initiatives began managed to collect less than ten per cent of the estimated wastes produced in them (Agyemang et al. 1997; Doe and Tetteh 1999; Stren 1989a). This means that as much as 90 per cent of these cities' solid waste products were either deposited in or near the residential areas that produced them, or burned in those neighborhoods. In all cases, whatever proportion the collected wastes comprised, they were deposited in uncontrolled dumpsites – some in or near the city centers, and others on the ever-expanding peripheries. Taken together, the collection and deposition crises can be seen as conducive to a plethora of environmental disasters. Leachate from uncontrolled dumps and local pits has created substantial water pollution problems, where burning produces localized air pollution concerns. Solid waste problems exacerbate chronic flooding in marginal urban communities and literally feed into vermin population explosions that increase the deterioration of human health (McGranahan and Satterthwaite 2002; Drakakis-Smith 1995). In city after city, the SCP branches concluded that some solution to the solid waste crisis would need to be prioritized.

Tanzanian and Zambian Cases

The choice of the cities in Tanzania and Zambia for the case studies is made for several reasons, beyond my own greater familiarity with them. First, it makes sense for the two Tanzanian cases to comprise much of the book on multiple counts (see Figure 3.2). Tanzania has been among the top three countries in the world in annual urban growth for four decades. Dar es Salaam, by many estimates, has been among the most rapidly growing major cities in the world; it is also the pilot city for the Sustainable Cities Program (see Figure 3.1). Although Dar es Salaam's dominance of Tanzania's urban hierarchy has generally increased with time, it hasn't been the only very rapidly growing city in Tanzania. In fact, one subtle trend of Africa's rapid urban growth has been the expansion of secondary cities (see Table 2.1). Partially in recognition of this, Tanzania was the world's first country to attempt to spread the ideas behind the Sustainable Cities Program to its entire urban hierarchy and to make that program's agenda the primary basis of urban planning in the country. At the level of a national hierarchy of cities, Tanzania's rhetorical and applied commitment to the SCP far outstrips the rest of the world. Zanzibar has had the longest experience with attempting to replicate the Dar es Salaam pilot project in Tanzania. Although Zanzibar is actually the country's third- or sixth-largest city (depending on how the urban population data is tabulated), its political status as a secondary capital for the semi-autonomous polity also called Zanzibar makes it Tanzania's second city, and thus the ideal second case study (see Figure 4.2).

Tanzania provided Africa and the world with a major model of alternative development thinking from 1967–1985. Under the banner of the Arusha Declaration and the intellectual and philosophical leadership of Julius Nyerere (1972), Tanzania's *ujamaa* [family-ness] policies sought an entirely different form of "decentralization" than that of today's development language. Tanzania's socialist leanings – a little more firmly to the authoritarian state socialist side in Zanzibar – and alternative idealism set it apart from the western world order. That idealized delinking came crashing down in the second half of the 1980s as Tanzania signed on to a program of structural adjustment. In the two decades since then, Tanzania has gradually evolved

Table 2.1   Tanzania's urban hierarchy (urban areas over 100,000 in population)

CITY NAME	2002 Population	1988 Population
Dar es Salaam*	2,497,940	1,360,850
Mwanza*	476,646	223,013
Zanzibar*	391,002	208,137
Dodoma*	324,347	203,833
Arusha	282,712	134,708
Mbeya	266,422	152,844
Tanga	243,580	187,155
Morogoro	228,863	117,760
Tabora	188,808	93,506
Sumbawanga	147,483	91,972
Kigoma*	144,852	84,647
Moshi	**144,336**	**96,838**
Shinyanga	135,166	100,724
Songea	131,336	86,880
Singida	115,354	80,987
Musoma	108,242	68,536
Iringa	**106,668**	**84,860**

Source: Bureau of Statistics, United Republic of Tanzania
* Indicates metropolitan area populations.
Boldface Type indicates participant cities in the Sustainable Cities Program.

toward model-state status in the IFI universe. It has assertively pursued neoliberal economic policies, prominently displayed its rhetorical commitment to the environment and sustainable development from a neoliberal perspective, and made good governance such a linchpin of policy that a Minister for Good Governance serves in the President's cabinet. Although Tanzania has not faced the traumas of civil war directly, it is a site of substantial significance for cultural politics in contemporary times – as host nation to millions of refugees from neighboring wars, and the site of trials for the genocide in Rwanda in 1994. The bombing of the US embassy in Dar es Salaam in 1998 underscored Tanzania's significance as a site in the global conflicts with – and within – Islamism. Perhaps most significantly for this book, racial, ethnic, religious and regional tension come together in the long simmering tensions between mainland and island Tanzania, and between the peoples of the two islands of Zanzibar. For all of these reasons, then – its particular experiences with neoliberalism, sustainable development (through its prominence in the SCP), good governance, and the politics of cultural difference – Tanzania makes for an extremely intriguing place to start the case studies, in chapters three and four.

The framework of the Sustainable Cities Program has been spread widely in eastern and southern Africa. Lusaka was one of the earliest African cities outside of Tanzania to send representatives to Dar es Salaam to learn about the program. Zambia has become one of Sub-Saharan Africa's most urbanized societies, and Lusaka has become its chief urban area after decades in the shadow of the Copperbelt mining cities in the north of the country. It was also for quite some time among the most active of Demonstration Cities in the world in the wake of the Dar es Salaam pilot project.

Like Tanzania, Zambia also embarked on an alternative development trajectory soon after independence. The Humanist philosophy of Kenneth Kaunda, Zambia's President from independence until 1991, mirrored the *ujamaa* philosophy of Tanzania's Nyerere (Meebelo 1973). Both are often considered to be forms of African socialism. But just as Nyerere relied on consultants from the capitalist West and northern Europe for urban planning, so Kaunda relied – rather heavily, in fact – on the World Bank. Lusaka was home to a major urban upgrading scheme with World Bank funding in the 1970s that attempted to establish entirely new neighborhoods while remaking much of the existing cityscape (Rakodi 1986b). By the end of Kaunda's time, though, Zambia was broke and in debt, and the city services that in its early days his regime had attempted to build up – or provide for the very first time – had generally fallen apart. His immediate successor, Frederick Chiluba, placed Zambia much more firmly in the grip of neoliberal policy, and also ushered in the Sustainable Cities Program (Rakner 2003a; Saasa 2002). Chiluba's successor, Levy Mwanawasa, has since 2001 brought a strong dose of good governance rhetoric into the mix, with his prominent engagement against corruption (even to the point of arresting and detaining Chiluba). Zambia, like Tanzania, has been relatively stable and at peace since its independence, but the politics of cultural difference bubble below the surface, in racial, religious, regional, gendered, and ethnic terms. The powerful hand of white, Colored, and Asian minorities and expatriates on the economy, and the mismatch of prominent Christian evangelical fervor with political leadership and with the practicalities of Zambian everyday life are hard to miss in the early 21st century. The stresses are showing from the inheritance of British colonialism's cobbling together of Barotseland (Western Province plus much of Southern and Northwest Provinces) and Northeast Rhodesia (with its strong Bemba/ Northern and Chewa/Eastern culture areas). Zambia after 1991 virtually flew into a freewheeling spiral downwards economically, and its failure to pull out of that spiral has exacerbated these other tensions. All of these factors make Lusaka, as a central test of the applicability of the Dar es Salaam pilot program of the SCP elsewhere in Africa, an important case study (see Figure 5.1).

The whole process of sustainable cities planning under the UN banner began in Dar es Salaam, which is the subject of chapter three. It rapidly spread from there, as a constellation of ideas and as a planning agenda, to dozens of African cities, including Zanzibar and Lusaka where we travel in chapters four and five. These chapters suggest that a majority of Zambian and Tanzanian planners and officials associated with the entire field of approach the SCP represents – particularly in waste management – expressed cautious optimism about it as a democratic, participatory, and efficient improvement on the past in interviews with me in 2002 and 2003. Yet what about the people being planned, or, apparently, empowered? As part of the process of analyzing the new agenda, I set out to enjoin assessments of the SCP in solid waste management – and urban management more generally – by the people carrying out the policies with the analyses of ordinary residents. A much more critical assessment results, in the chapters that follow this one.

In the section below, I place the SCP in its larger context of the UN system and detail what it is that the SCP actually sets out to do. I am interrogating how the SCP navigates the discursive landscape of development in this era – how it engages with neoliberalism, sustainable development, good governance, and the politics of cultural difference in African cities.

The United Nations and African Urban Development

The SCP is one of 19 different programs under the UN-Habitat umbrella. These programs blend into one another and, in cities that participate in more than one, they may even have overlapping personnel. The list itself gives away much of the rhetorical spin to Habitat's agenda. There are programs for Best Practices and Local Leadership, Cities without Slums, Gender Policy, Global Urban Statistics, Housing Policy, and Housing Rights. Other offices focus on Urban Poverty, Land Tenure, Managing Water for African Cities, Risk and Disaster Management, Safer Cities, Training and Capacity Building, Urban Transport, Urban Sanitation and Solid Waste Management, and Urban Economy and Finance. Habitat runs an Urban Environment Forum, a program for Localizing Agenda 21, a Campaign on Urban Governance, and the Global Campaign on Secure Tenure, on top of coordinating the Urban Management Program with UN Development Program and the World Bank. Ambitious, practical, and progressive phraseology abounds – who could argue against the need for better managed, safer cities with local leadership and without slums?

The SCP is only one part of a vast and diverse fleet that – at the rhetorical level – seems to be fishing in the right waters. The whole fleet is loosely tied to a framework for environmental sustainability and good governance that goes under a different label than Agenda 21, using the acronym, EPM, for Environmental Planning and Management. The EPM framework, like most urban planning models around the world today, emphasizes the planning roles of private sector and civic organization stakeholders in partnerships with the state in a decentralization of market-efficient decision-making (Mohan and Stokke 2000; Beall 2002). EPM and the broader agenda with which it is associated has sought to transform urban development planning and policy in much of Africa. The Sustainable Cities Program describes its global goals for EPM as working "towards the development of a sustainable urban environment, building capacity for urban environmental planning and management, and promoting a broad-based participatory planning process" (SCP 2004, p. 1). Really, EPM is more of a program with a "gospel" for re-engineering the processes of urban development and governance than one for improving the environment in cities (Kitilla 2001, p. 86 and Kitilla 2003; Halla and Majani 1999a). The Tanzanian planner, Martin Kitilla, who utilized the term, gospel, to describe the EPM framework, was not really exaggerating: it is a framework for which the UN has proselytized across the continent and the globe. EPM and its buzzwords – participation, capacity building, partnerships, enablement, privatization, and decentralization – run through all nineteen Habitat programs in one form or another, so it is useful to spend some time understanding it (Dahiya and Pugh 2000, p. 157).

The EPM process has a neat flow chart that begins with the compilation of an environmental profile by city governments. This entails creation of a thorough inventory of resources, problem areas, socio-economic trends, and policy dynamics by a team of planners. The next step is termed a city consultation, in which a wide variety of stakeholders are called together to develop a consensus on which environmental planning problems to prioritize, using the disseminated profile and other commissioned working papers as a starting point. The EPM process calls for the use of four criteria for defining what stakeholders are invited to the consultation. Essentially these boil down to those who cause the problems, those who are affected by them, those who have the political power, and those who have the means to help solve them (Mlambo 2003; Dahiya and Pugh 2000). The list of those invited and the range of those who attend can do much to determine the actual inclusiveness of the process, and it has essential impacts on outcomes. This is first and foremost because it is out of the city consultation that working groups of stakeholders are formed to tackle specific environmental planning problems, and then the stakeholder working groups are supposed to conduct mini-consultations to devise action plans for pilot areas. Typically, this begins with a "demonstration project" to instill a sense of "ownership" of the program in the public (Dahiya and Pugh 2000, p. 161). Public-private-popular partnerships, manifested in the working groups, are then supposed to seek funding or capital to implement the action plans (see Figure 2.2).

Ultimately, the EPM process is supposed to emerge from these diffuse, sectorally focused, and geographically circumscribed action plans with a strategic urban development plan for each participant city (Halla and Majani 1999a and 1999b; Halla 1994; UNCHS 1992; Dar es Salaam City Commission 1999). In its flexibility, transparency, fluidity, and democracy, the EPM process seeks to improve upon and supplant the top-down blueprint master planning that previously predominated in participant cities.

This EPM framework's rhetorical ideals include a familiar combination – liberalized economies, ostensibly democratic and accountable politics and governance, and environmental sustainability (Parenteau 1996; Sandilands 1996). Like other Habitat or UNEP programs, the SCP works through and depends upon various tiers of bureaucratic power, from the global level of the UN bureaucracy through national and municipal governments to neighborhoods, wards, and households (SCP 2000 and 2001). In fact, the UN parent office claims to want to work the other way round, from the bottom up (Eigen 1998; Desai 1999). The SCP seeks to manufacture a kind of "deep democracy" (Appadurai 2001) or at the very least a "stakeholder democracy" (Freund 2001). With the SCP, as with many developed and developing world initiatives like this, planning and development are conceived of as collaborative visioning among a broad array of stakeholders who in the process are alleged to be

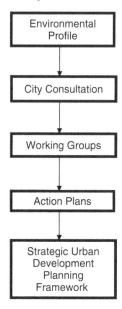

Figure 2.2   The EPM gospel
Source: author

"producing new geographies of governmentality" (Appadurai 2001, p. 25; see also McCann 2001).

The rhetoric of the SCP often seems to borrow from what has been called the "argumentative turn" in urban planning, toward more deliberative processes (Fischer and Forester 1993). This deliberative planning "inspires a collaborative approach to decision-making in which citizens come together to develop collective responses" (Hayward 2003, pp. 114–115). The Tanzanian economist, Anna Tibaijuka, cited at the head of this chapter, who is Habitat's Executive Director and served as a member of British Prime Minister Tony Blair's blue-ribbon Commission for Africa in 2004–05, sees the SCP as "one of the flagship products of the global campaign on urban governance" (UNCHS 2001, p. i). To repeat the quotation with which the chapter begins, the chief theme of this campaign "is to help realize the 'Inclusive City' – a place where everyone, regardless of wealth, gender, age, race or religion, is enabled to participate productively and positively in the opportunities that cities have to offer" (UNCHS 2001, p. i). Arguably, the most famous "inclusive" ideal is that of sustainability, since local stakeholders are seen as the defining agents for it in particular circumstances. The United Nations report on "An Urbanizing World" defined sustainability by its now famous phrase: "meeting the needs of the present ... without compromising the ability of future generations to meet their own needs." Concerns for "socio-economic development" are supposed to be balanced with those for "global ecological life-supporting systems" for all parts of the earth. But the political paths of policies interrupt the flow toward that balance, particularly in the many African contexts that lack the transparent, accountable, and participatory political systems that neoliberals claim to desire.

Indeed, the collaborative visioning within the SCP seeks to localize global initiatives that are full of neoliberalism with a human face, even while sidestepping thorny realities (Mercer 1999; Burgess, Carmona, and Kolstee 1997). The UN's SCP advocates see it as a program to improve the effectiveness and efficiency of inter-agency coordination within states during an era of state withdrawal from the realms of collective consumption and urban service provision (Eigen 1998). But there are some crucial *a priori* assumptions: state withdrawal is a good thing, an immediate reliance on external assistance is the solution to financing action plans, and decentralization can be accepted uncritically. This acceptance comes without recognition of how politicized the processes are, particularly in the highly charged contemporary politics of cultural difference.

There are a number of Habitat programs that, on paper, are a little broader than a straightforwardly capitalist, western-world political agenda. The "urban" needs of the present that are generally included within the sustainability language of Habitat documents (such as UNCHS 1998, p. 1, or UNCHS 1996, pp. 6–10) comprise an interesting list. Typically, access to urban services, health care, and the "freedom to participate in national and local decision-making processes" take rhetorical priority. It is not hard to agree with Kipfer (1996, p. 122) that "the concept of urban sustainability" that Habitat embraces "seems to acknowledge at least rhetorically the need for institutional change and the inclusion of non-capitalist criteria into the planning process." Although a genuine "democratization of the state" might flow from the implementation of the concept, "the possibility of deepening the discourse of 'urban sustainability' depends on the particular balance of socio-political forces which must be determined in each concrete case" (Kipfer 1996, p. 122). The various Habitat programs to some degree leave open the possibility for developing a balance of socio-political forces in the progressive interests of the urban majority if the particular cultural politics of a place allows for it. But they equally can be construed in ways that simply maintain the status quo. And the latter is, almost always, the well traveled path, as the case study chapters show.

In sum, the ideas sound charming. As Drakakis-Smith (1995, p. 672) put it, though, the UN's "progressive documents ... flatter to deceive." It is increasingly apparent that sustainable development "has become a convenient slogan to signal political correctness without the corresponding commitment to change" (Simon 1999a, p. 28). Behind the rhetoric, the same modernization or neoliberal agenda of "economic efficiency" – or something eerily similar to it – still appears to dominate, and genuinely "people-centered and indigenously generated African alternatives" seem to get the short end of the stick (Simon 1999a, p. 29 and 1995a, p. 36). The 2002 World Summit on Sustainable Development (WSSD) in Johannesburg extended the neoliberal agenda still further. In the WSSD declaration and its plan of implementation, free markets and private investments achieved even greater prominence as the solutions to nearly every human settlements issue, and local cultures and politics were further shunted aside (Middleton and O'Keefe 2003).

Even if we give the UN the benefit of the doubt for having the best intentions of the ordinary poor majority at heart, there are still at least two gleaming barriers that block the path to the Inclusive City in Africa. The first is donor dependence, and the second is the generally combative and exclusive character of local urban settings of implementation. I examine each in turn in two sections below. I spend much more time on the former, because the latter is a main point of the case studies in chapters three to five.

Top-down Donor-driven Bottom-up Localized Development

Much of the funding for implementation of SCP programs across Africa comes from northern European donors. The SCP's work in Tanzania alone has received funding, in various action plans or visioning schemes, from Denmark, Germany, Sweden, Britain, Finland, Norway, and France. The Danish, German, Swedish, British, and Norwegian aid agencies have also funded some aspects of the efforts in Lusaka. They have been joined in Zambia by the aid agencies of the U.S., Japan, Ireland, and Canada (see Table 2.2). This means SCP funding has come mainly from six of the eight so-called G-8 countries

Table 2.2   Donors to the SCP, case study cities

AGENCY	CITIES FUNDED
DIRECT FUNDING FOR SCP PROGRAM ACTIVITIES	
Danida (Danish International Development Agency)	Dar es Salaam, Lusaka
DED (German Development Service)	Dar es Salaam, Zanzibar
GTZ (German Society for Technical Cooperation)	Dar es Salaam
ILO (International Labor Organization)	Dar es Salaam, Lusaka
Ireland Aid	Dar es Salaam, Lusaka
KFW (German Development Bank)	Zanzibar
SIDA (Swedish International Development Agency)	Lusaka
UNCHS (United Nations Center for Human Settlements) (Habitat)	All three cities
UNDP (United Nations Development Program)	All three cities
UNEP (United Nations Environment Program)	All three cities
OTHER URBAN DEVELOPMENT FUNDING RELEVANT TO SCP WORK	
DFID (UK Department for International Development)	Lusaka
Finnida (Finnish International Development Agency)	Lusaka, Zanzibar
JICA (Japan International Cooperative Agency)	Lusaka
KFW	Lusaka
NORAD (Norwegian Agency for Development)	Lusaka
SIDA	Dar es Salaam, Zanzibar

(only Italy and Russia are absent here) and 11 of the richest countries in the Organization for Economic Cooperation and Development (OECD). Several international non-governmental organizations and private charitable foundations – such as CARE and the Ford Foundation – have played important roles in pilot schemes. The net total funding provided by this alphabet soup of donors to SCP-related programs across Africa is now above fifty million US dollars. The G-8 and its OECD friends in high (latitude) places must want something to work from the millions that have been poured into African cities. Is that something the smooth removal of barriers to their further investments in these countries, the reconstitution of governance to more effectively regulate the dangerous corners of their periphery, or an improved quality of life and living environment for the world's poorest urban residents? There is probably something for everybody amongst these. But what do these dollars amount to for the poor majority, and why do these donors care?

Among these countries, the US spends the lowest percentage of its national budget on development aid, sends the lowest portion of its aid budget to Africa, and has the least to do with funding actual sustainable cities programming. Canada, Norway, Japan, Britain, Finland, and France have been involved only indirectly in funding sustainable cities activities, particularly in the three case study settings. For instance, in Lusaka over the last few years, Britain (through its Department for International Development, DFID) has funded a water and sanitation project in the Mtendere area. Japan (via the Japan International Cooperative Agency, JICA) has built roads and water connections throughout the city. Norway (whose agency goes by NORAD) has funded the construction of a community center and clinic in Ng'ombe. None of these projects are tied directly to the Sustainable Cities Program, nor has the EPM framework played any role in them. Given my purposes in investigating how and why aid flows into SCP/EPM schemes, I concentrate on German, Irish, Swedish, and Danish aid here.

German aid has flowed into community upgrading and solid waste management schemes in Zanzibar and Dar es Salaam. Various German donor agencies, including GTZ (the main German aid agency, *Gesellschaft fur Technische Zusammenarbeit* – Technical Cooperation Society – which is basically a government-run corporation), KFW (*Kreditanstalt fur Wiederaufbau*, the German Development Bank), and the DED (*Deutscher Entwicklungsdienst*, or German Development Service), have funded SCP projects. German aid has been most substantial in Zanzibar, including the money and technical assistance to the Mkele pilot program for solid waste management discussed in chapter one (and chapter four). German aid also backed a pilot program for community consultation and public-private partnership in Dar es Salaam's sustainable program.

German aid rhetoric is the most straightforwardly capitalist of the four. Germany is also the one former player of these four in the African game of colonialism, since Tanganyika (mainland Tanzania) was a German colony from the 1880s until the end of the First World War and Germany had designs on Zanzibar in the nineteenth century. Although those colonial ambitions are, at least explicitly, confined to the distant past, it is hard not to note the paternalistic condescension in German government analyses of Tanzanian development in the present. The German embassy's website (http://www.german-embassy-daressalaam.de), for instance, praises Tanzania's shift toward "a more pluralistic environment" politically and a "market oriented economy," yet it is blunt in its critique of the shortcomings of this transition:

> Although the business and investment climate has improved, outdated and inconsistent company legislation, burdensome tax administration, red tape and a weak judiciary still pose significant obstacles for the private sector. The quality of public services is generally poor, financial management and accountability continues to be weak thus increasing opportunities for corrupt practices.

German aid is thus seen as an antidote to the bad behavior of a Tanzanian state still suffering from a socialist hangover. The colonial past is not the only glaring silence here. Communist East Germany's important history of urban development assistance to that socialist system (in Zanzibar in particular, where planners from the DDR wrote the city's 1968 master plan) is totally absent from the new Germany's recollection of Tanzanian-German cooperation (http://www.gtz.de; Myers 1994b). Instead, the rhetoric here matches directly with that of the IFIs – Germany seeks to use aid to strengthen local governance for capitalist development, full stop. A similar rubric pertains in German government discussion of its much less significant aid flows to Zambia.

Tanzania and Zambia are among the four longstanding priority countries for Irish Aid in the world (Lesotho and Sudan are the others). Almost all of Ireland's bilateral (government to government) aid has gone to rural development projects, with the significant exception of its support for sustainable urban planning work in Lusaka. Irish Aid has been a substantial contributor to the Sustainable Cities Program in Lusaka, through its community upgrading programs in several pilot areas for the SCP, in collaboration with CARE. As a much more small-scale and low-key donor, and as an OECD country long colonized by the same power that once ruled Tanzania and Zambia (Britain), in certain respects Ireland proves a more sympathetic partner than Germany for implementing what is supposed to be a more populist development program. A March 2002 government review of overseas development assistance stressed that "Ireland has a long tradition of solidarity with the poor and the dispossessed. Our own history of colonization, poverty, famine, and mass emigration, and our commitment to a fair and just world order predispose us to this" (Ireland Aid Review Committee 2002, p. 2). The Irish public has one of the world's highest rates of voluntary contribution to humanitarian assistance and other "Third World causes" (Ireland Aid Review Committee 2002, p. 2).

Ireland's government, although a practitioner of neoliberalism and good governance at home, is not explicitly strident in demanding these of its aid recipients. Ireland Aid stresses "direct involvement with local communities, and environmental and gender concerns are given a high priority." All aid to all countries has long been "untied," meaning it does not depend on tie-ins to the use of Irish firms or services. At a time that many OECD countries have substantially reduced their foreign aid budgets, Ireland has done exactly the reverse, by attempting to fund steep annual increases in assistance. Nevertheless, the March 2002 review stresses "a new policy focus" on "governance, democracy and human rights" alongside "sustainable development" and a "support for the private sector in developing countries" (http://www.ireland-information.com). The World Bank's assessment of Ireland Aid's work in Lusaka (World Bank 2002, p. 5) has in common with the German rhetoric above a tendency to blame "the years of central planning" in postcolonial Zambia (1973–1991) for "the development of a culture of dependence on the state and the top-down provision of services, which has resulted in citizens not expecting or wanting to pay for services enjoyed or consumed." This critique is stunningly absent of an appreciation of British colonial legacies in creating these problems, to the extent that they may exist at all. In the end, the country assessment concludes that Zambian government policies on housing and urban services "should be reviewed with an eye toward allowing for private sector competition in the supply of goods and services" (World Bank 2002, p. 20). This is, of course, the World Bank's interpretation, but it carries the Ireland Aid stamp of approval. Despite a more realistic and sympathetic assessment of the politics of decentralizing urban planning to the community level and progressive nods toward human rights or gender equity, the Irish Aid programs thus appear to be quite in line with the IFI approach to development.

Sweden and especially Denmark – largely through SIDA and Danida, the Swedish and Danish International Development Agencies – have been major financial supporters for SCP activities in Tanzania, Zambia, and the whole of eastern and southern Africa, and therefore deserve more scrutiny. They also give far greater proportions of their GNP to development aid than Germany or Ireland, consistently placing them among the top countries in the world in net bilateral development assistance as a percentage of GNP (Potter et al. 1999). The Nordic countries (Sweden, Norway, Finland, and Denmark), and Sweden especially, have a deep association with UN activities in southern Africa, and throughout much of the 1980s and 1990s they provided nearly 30 per cent of all OECD aid to the region (Sellstrom 1999a, pp. 50–55). Both countries have a long history of aid tied to what are, rhetorically at least, progressive political agendas – Sweden was the first western country to impose sanctions on apartheid South Africa, for instance. Whether this history and pattern of generous assistance shapes the SCP or EPM in more progressive directions than the German or Irish aid is another matter.

SIDA has funded environmental management information systems (EMIS) programs, community-based infrastructure rehabilitation schemes, land management programs, and/or urban governance projects in all of the case study cities. Swedish aid to the SCP and the implementation of the EPM framework is in keeping with a longstanding Swedish rhetorical commitment to popular participation in development (Rosander 1992). Growth, equality, democracy, socio-economic independence and sustainability have been Sweden's official goals for aid to African countries for twenty-five years. Given Tanzania's commitment to non-alignment and its expressed policy of

ujamaa and African socialism for much of the post-colonial period (1967–
1985), it is probably unsurprising to find historically democratic-socialist
Sweden atop the list of long-term donors to the country. Zambia's similar
foreign policies and domestic development commitments in the early
postcolonial period also attracted substantial Swedish aid. Human rights and
social welfare for the poor have been central to these aid programs for nearly
four decades. The connections are so strong in Tanzania that the departure of
Sweden's ambassador, Sten Rylander, in June 2003 – normally a perfunctory
matter of minor significance – was a leading news story for the whole month
beforehand. Lengthy newspaper or magazine features and radio shows
interviewing both Rylander and all sorts of Tanzanians suggested a deep and
broad affection, not just with this man, but with the tradition of aid and
support, among Tanzanians.

Yet all is not as rosy as it seemed in the radio interviews. The problems are
with both rhetoric and substance. Eva Rosander (1992, p. 37) has shown that
persuasion and exhortation are the emotional foundations of Swedish rhetoric
on "people's participation." The different personnel involved in carrying out
the policies attached to that rhetoric have very different interpretations of its
practical implications and highly varied levels of commitment to the call to
persuade and exhort. On one hand, there can be a smug assumption of the
superiority of Swedish values and ideals that leads to their forceful promotion,
especially by those farthest away from actual day-to-day project implementation. Rosander (1992, p. 60) found that in Swedish development programs,
"participatory efforts contained an element of 'cognitive imperialism.' The
Swedes imposed their own particular modes of perception, evaluation and
action on [people] ... who organized their relationship to reality differently."
On the other hand, SIDA field practitioners have tended to be far less
committed to the rhetorical ideals, dismissing them as "too theoretical"
(Rosander 1992, p. 47). That can mean, in practice, little regard is taken for
popular participation, or other progressive ideals, in implementation.

Beyond rhetorical ideals, SIDA programs in Tanzania and Zambia are
increasingly indistinguishable from those of other OECD donors or the IFIs.
In Tanzania, SIDA supports the "privatization of inefficient state companies"
and "policies facilitating free enterprise." It calls corruption "one of
Tanzania's Achilles heels." Despite saying that "Tanzania is responsible for
its own development and decides what initiatives should be prioritized," the
document that SIDA sees as guiding its aid is the Poverty Reduction Strategy
Paper that Tanzania was forced to create as a condition of its aid and debt
relief agreements with the IFIs (http://www.sida.se).

Similar contradictions and conundrums are apparent in Danish aid. Danida
has funded participatory solid waste management and programs for the
privatization of solid waste management in several of Tanzania's SCP cities
(including Moshi, Mwanza, Arusha, Iringa, and Tanga) as well as Lusaka.
Tanzania has received more Danish aid than any other country every year since
1997, while Zambia has been in the top ten among countries receiving Danida
funds (it is placed in the second tier of priority countries, while Tanzania tops
the first tier). Environmentalism, human rights, gender equity, and "pro-poor"

policies are hallmarks of Danish assistance to these and other countries (Danish Ministry of Foreign Affairs 2003). Denmark has also focused considerable efforts on "redefining development" to account for "cultural insight into broad development strategies." Cultural development – "strengthening the cultural sector" – for Danida (2002a) means investing in "mainstreaming culture in all development work," using "culture for development" projects, and "analyzing the consequences of development cooperation on the culture of a country, community, or group." The Danes see themselves as leaders in recognizing the politics of cultural difference that impact and are impacted by development processes. Even more than in German, Irish, or Swedish aid documents, one can read the explicit ways in which Danish policies changed in response to the events in the U.S. on September 11, 2001. For instance, Denmark pledged to respond to the heightened tension in the world by "tackling political radicalism and religious fundamentalism," cooperating "more closely with those groups in the Arab world supporting modernization and democratization" (Danida 2003). Although Zanzibar is not a part of the Arab world, it is 95 per cent Muslim, and it is striking to note that Zanzibar city is the only Tanzanian participant city in the sustainable cities network to not receive aid from Danida.

As Danish domestic policies drifted toward the political center during the 1990s and 2000s, so too the associated discourse on international development. Denmark's total net overseas development assistance declined in real terms and as a percentage of the GNP. Denmark's policy statements began to sound identical to the IFIs – to the point of citing World Bank President James Wolfensohn as one of the leading advocates of infusing culture into development in their document on *The Power of Culture* (Danida 2002a). Danida's *Country Strategy for Tanzania 2001–2005* promoted poverty reduction, sustainable development, good governance and partnership "with a variety of stakeholders, such as the government, the private sector, and the civil society" (Danida 2002b). Their goal was a "market-based economy where the private sector is expected to play a leading role." Support for "the business sector" became the first priority. In early 2002, Denmark's government completely reevaluated Danida's mission, and the resulting re-orientation only strengthened the neoliberal bent toward sustainable development and good governance. Danish assistance was to be first and foremost an effort to "support the development of a private sector as an engine for growth." Accordingly, the Danes now said that they would reduce environmental assistance to Tanzania "to ensure the necessary ownership of future environmental efforts" (Danida 2003).

As Marcus Power (2003, p. 135) succinctly puts it, "progressive intentions do not always translate easily into development practice." Northern European aid programs often have offered paeans to gender equity, human rights, social equality, popular participation, environmental justice, and the like. It must be said that certain programs most assuredly worked in that direction on the ground in Tanzania (and to a lesser extent in Zambia) in the 1960s, 1970s, and early 1980s. Whatever progressive leanings remain in Irish, Swedish, or Danish aid, the predominant language is the language of neoliberalism's version of sustainable development and good governance. With Denmark's nods to *The Power of Culture* aside, there appears to be little recognition of how the local politics of cultural difference shape development outcomes, of how to work with those local particularities, or of why that might matter.

It is in that context of aid rhetoric that the aid from these countries operates in the sustainable cities program. Since these countries have made it abundantly clear that they will pull aid from countries whose policies do not meet their criteria – all four froze aid to Zanzibar after its questionable 1995 elections, for example – we end up with a truly weird circumstance. The Sustainable Cities Program's EPM framework sells itself as being about bottom-up localized urban management led by working group coalitions of stakeholders. Presumably, the donor countries I have been examining sign up to support the EPM's application because they are inspired by its bottom-up, localized character. Yet one cannot help but notice that the EPM framework fits all of the demands of the donors who support it. The donors' agents often operate to shape the directions the sustainable cities program offices take on a day to day basis. Hence the weird result: a top-down donor driven bottom-up localized development initiative. And it seems almost self-evident that "healthy and secure environments for the urban poor ... cannot be imposed top-down" (Dorman 2002, p. 146).

The Exclusive City

In addition to the shaping role of donors with their own particular agendas that do not necessarily match up with progressive ideals, "localized conditions of leadership, policy reform, and institutional and cultural settings" can be trump cards for the outcomes in specific cities (Dahiya and Pugh 2000, p. 163). The global SCP office wants to produce a defined set of interrelationships through which agencies interact efficiently. But in operationalizing this, the African SCP offices meet with urban majorities that are already excluded, disengaged and cast out by processes begun long before the SCP came along (Simone 2001a and 2001b). In Africa's SCP cities, the excluded majorities are instead in motion reconstructing their neighborhoods from below. As Arturo Escobar put it in a Central American context, "local groups, far from being passive receivers of transnational conditions, actively shape the process of constructing identities, social relations, and economic practice" (Escobar 2001, p. 155). But they do so outside of the UN's feedback loops or the donors' gaze, and within the continuing constraints of state repression and global economic structures and flows that derail the possibilities that any new planning gospel might offer them. What, for instance, might EPM have to offer the teams of Dar es Salaam squatters who dig potholes in the streets of rich neighborhoods and then seek payment from passing motorists for filling them?

The politics of cultural difference in an exclusionary democracy character-ized by such a class divide cancel out the EPM gospel, or at least its progressive possibilities. The "trivialization of history" and "assumed harmony among and within social forces" embedded in the EPM gospel thoroughly doom it

(Ihonvbere 1996, p. 6). The "character and results" of Sustainable Cities activities "are determined by relations of power" that work entirely against progressive instincts (Kothari and Minogue 2002, p. 13). This argument becomes clear in each of the three case study chapters, but let me spell out its essence here briefly.

First, the "acute service deprivation" associated with neoliberal structural adjustment in African cities since the mid-1980s is often said to have encouraged Africans to "devise alternative modes of survival" (Halfani 1996, p. 87). As Mhamba and Titus (2001, p. 227) put it, "the overwhelming reaction of people in the city to escalating urban problems can be said to be that of performing operations themselves, either individually or collectively." Ordinary people in Sub-Saharan African cities, though, are quite practiced at doing for themselves. In most cities in British settler and labor-reserve colonies, for instance, Africans could not even legally be in cities without passes around their necks. Because so very many people did come to live in and around cities – particularly in the late colonial era – in marginal zones, in fear of authority, "performing operations themselves" in the city was the only means of assuring these operations would be performed at all. This includes operations like food production, since urban and peri-urban agriculture has been a mainstay of the urban poor for generations. It also includes devising ways of dealing with solid waste. As these cities have expanded, the patterns of "doing for oneself" have expanded apace. Asking for, or, worse yet, assuming a collective relationship with urban authorities or a mythical urban private sector that might undergird a "stakeholder" democracy is a pie-in-the-sky fantasy when one thinks of how wide the gulf has been between the states or elites on one hand and the urban poor majority on the other in these cities.

Secondly, this assumption is made even more fantastic by the processes of imposition of multi-party democracy and good governance in cities where cultural differences have been deliberately utilized for political purposes for a century. Local politics in many African cities are so combative because they were set up to be that way. British colonialism consciously shortchanged the development of local authorities and shaped them to serve its purposes. Underfunded, racially divided, with class poised against class, ethnic group against ethnic group, municipal councils of the late colonial era usually bore no resemblance to the cities they supposedly represented, and real authority rested higher up in the hierarchy. Postcolonial regimes usually discarded the racial codes that governed such councils, but they often discarded the councils, too: Tanzania, Zanzibar, and Zambia, for instance, had no local governmental bodies for significant stretches of the postcolonial era. People have been born, grown up and died without seeing a functioning municipal democracy engaged in any partnerships worth a damn in the provision of urban services.

Moreover, the new era of democracy is one in which central governments, ruling parties, IFIs, and donors often act in concert to restrict the legitimacy of political parties that represent the poor (Abrahamsen 2000, p. 82), as in the Democratic Party (DP) or the Civic United Front (CUF) in Tanzania. Donors and the IFIs get used to working with the rulers. Perhaps they might find it hard to have their way quite so much with a President (Reverend) Christopher

Mtikila (of DP) or (Maalim) Seif Shariff Hamad (of CUF), both of whom have demonstrated no shortage of rhetorical fire and brimstone religiosity against the establishment. Yes, the donors froze aid to Zanzibar when Hamad was defeated in the questionable 1995 elections. But they did not freeze aid to *Tanzania*, and looked the other way as Tanzanian – not Zanzibari – military and police imposed a brutal order. Donors made little or no note of the deregistration of Mtikila's party, which claimed to represent those that this outspoken Christian fundamentalist cleric labeled *walalahoi*, the dispossessed. In both instances, the donor's "partnership" with the Tanzanian government took on a new dimension, one that smacked much more of collusion. But this is an old collusion, really. It is colonialism's exclusionary democracy reborn, or revitalized. The response of the walalahoi is to shrug at their continued disenfranchisement and get on with the business of "performing operations themselves," even when that means dumping garbage in a drainage ditch, taking to the streets to reclaim a cemetery or refill a pothole they made the night before, or bombing a local leader's house.

Third, a great many of these "operations" are "damaging to the urban environment," making a mockery of the phrase, sustainable development (Halfani 1996, p. 87; Main and Williams 1994; Potts 1994). For instance, unregulated, high-density settlement in marginal areas often becomes the only way the poor have of coping with the out-of-control land markets and housing markets of a neoliberal order. In a privatized universe, such areas are unserviced or extremely underserviced. Their lack of wastewater or sanitation infrastructure or solid waste management systems can be associated with increased incidence of communicable or water-borne disease and a constella-tion of virulent forms of "environmental destruction" in urban Africa (Halfani 1996, pp. 87–88; Odumusu 2000). This expands the need not for the tepid forms of EPM that have been implemented, but for a genuine political ecology. By this I mean a "critical, historically informed analysis of both the emerging narratives and policies concerning the African urban environment and the concrete ways in which people experience environmental damage in urban areas" (Freidberg 2001, p. 349). The SCP is one of those "emerging narratives and policies," and popular negotiations with its aims are among the "concrete ways people experience environmental damage" in African cities. Hence, this is where I go in the case studies that follow.

Conclusion

During the past decade, a new agenda of good governance that combines neoliberal economics, multi-party politics, and urban environmentalism has come to eastern and southern Africa in force. One manifestation of this agenda has been the establishment of offices throughout the region for the United Nations Sustainable Cities Program. Through its highly-touted gospel of participatory planning, which it terms the Environmental Planning and Management (EPM) approach, the SCP has sought to empower the urban poor to provide their own services and build local economies in the process.

Like similarly populist rhetoric in neoliberal planning, this EPM approach has proven popular with development donors. The SCP has relied mainly on donors from northern Europe, and particularly several countries noted for their generosity towards eastern and southern Africa as well as their legacies of democratic socialism at home. Increasingly, though, these donors sound alike, and they blow the trumpets of the International Financial Institutions that call the discursive workhorses of the day to battle: neoliberalism, sustainable development, and good governance. The UN and donor programs, however, confront urban populations embroiled in a politics of cultural difference and seething with decades of resentment at their disenfranchisement in exclusionary democracies.

In the chapters that follow, I examine the planning processes that have devolved out of the SCP approach, with particular reference to solid waste management, in Dar es Salaam, Zanzibar, and Lusaka. I do so with a political-ecological eye on donor, state, and community actors and various scales. The mixed stew of neoliberalism, good governance and sustainable development rhetoric and action that SCP embodies professes to set about producing stakeholder democracies at the municipal level in cities like this across the continent. While progressive aspects remain lodged in various rhetorical corners of the new planning processes, the predominant result appears to be a reinforcement of exclusionary democracies that have been present in these cities since at least the late colonial era, in the midst of a politics of cultural difference that further disables any progressive possibilities.

Chapter 3

Dar es Salaam:
Model City for the World

We, residents and friends of this city, have one common vision ... (the) sustainable
city of Dar es Salaam ... a city where all residents have access to a decent livelihood.
(Dar es Salaam City Commission 1999, p. i)

Introduction

Tanzanian cities are, like many cities in Sub-Saharan Africa, undergoing a
process of profound transformation in economic, political, environmental, and
cultural life. Among the countries for which statistics are kept, Tanzania has
long languished in the bottom ten in the world in economic indicators. The
country has been one of the world's leading per-capita recipients of
development assistance for several decades. After nearly two decades of
socialist planning (1967–1985) that the IFIs blamed for enormous public debts,
Tanzania was forced into signing a structural adjustment plan in 1985, in effect
to pay back the previous decades of assistance. Structural adjustment
programming is credited by the IFIs with stabilizing Tanzania's economy by
providing "changes that had a positive bearing on private sector development"
(World Bank 2001, p. 63). The World Bank was so pleased that Tanzania
became, in 2001, the third African country to be offered debt relief under its
much heralded Heavily Indebted Poor Countries (HIPC) initiative. Most
scholarship on Tanzania's experience of structural adjustment, by Tanzanians
and others, paints a less rosy picture, of the negative, or even devastating
effects of SAP or HIPC on everyday survival for most Tanzanians, especially in
its largest city of Dar es Salaam (Lugalla 1995; Tripp 1997).

Tanzania has also switched from a single-party political system (in place
from 1967 to 1992) to a multi-party parliamentary democracy. Although the
multi-party era began in 1992, and November 1994 marked the dawn of multi-
party local councils, the October 1995 national presidential and parliamentary
elections are typically seen as its real beginning. Tanzanians sometimes describe
this change in terms of the Swahili proverb, "the donkey is the same; we have
only changed the pack-saddle," because of the ruling party's continued
dominance of a political-economic framework established from outside (Myers
1996). At independence in 1961, Tanganyika (mainland Tanzania) had a
government headed by the Tanganyika African National Union (TANU), the
only party in its parliament. Following Zanzibar's January 1964 revolution and

April 1964 unification with Tanganyika, the Afro-Shirazi Party (ASP) headed the islands' somewhat autonomous regime. Both parties ruled the respective wings as single-party states. In 1977, the TANU and ASP merged, under the direction of Tanzania's first president, Julius Nyerere, to form the Revolutionary Party. This ruling party is known by its Kiswahili acronym, CCM (Chama cha Mapinduzi, literally, Party of the Revolution). Despite the multi-party system that has been operative now through two national elections (1995 and 2000), and despite an evidently more open society in some ways, CCM has continued to attempt to control Tanzania's political stage and its government. It does so now firmly within the rubric provided for it by IFI neoliberal good governance, but CCM control rests atop a growing pile of local resentments. Those resentments, perhaps unsurprisingly, pile up rather high alongside the garbage in Dar es Salaam, as the country's largest city, economic engine, and political-cultural heart.

Tanzania's environments are among the most widely represented African landscapes in the world. They are also among the most formally protected, since more than 110,000 square kilometers of the land surface of the country is covered by a game reserve, national forest, national park, or conservation area. The Tanzanian government has developed a considerable array of institutions, laws, and policies since the 1990s dedicated to environmental planning, and sustainable development rhetoric is heavily utilized in government development documents (Bagachwa and Limbu 1995). Environmental programs are prominent in the priorities of the major donors to Tanzanian development, perhaps because of the visibility and representational character of the Tanzanian landscape (given the presence of Mount Kilimanjaro, the Serengeti Plain, and the islands of Zanzibar). Much of the extent of the country's resource richness has only been discovered in the past two decades, such as the Songosongo natural gas field, and the mineral wealth in gold, rubies, tanzanite, and other gems or precious metals. This new prominence to environmental planning more broadly provides yet another backdrop for why Dar es Salaam would serve as the pilot city for the Sustainable Cities Program.

Culturally, the last decade in particular has brought a whirlwind of change to Tanzania. The country's leadership once prided itself on an ideology that privileged self-reliance (*kujitegemea*), collective responsibility (*ujima*), social unity (*umoja*), family-ness (*ujamaa*), and similar virtues (these four – not coincidentally – having been inspirational in the creation of the African-American holiday of Kwanzaa). The old Tanzania promoted the Swahili language to unite its more than one hundred ethnic communities and to supplant the English language. The old Tanzania engaged in a lively, thoughtful, literate and very public debate over whether or not tourism was an industry worth investing in, given the demeaning character of so much of the employment it creates and the fact that so much of the income it generates remains with outsiders. The new Tanzania is a very different place, a land of *ubinafsishaji* (privatization). By the early 21st century, Tanzanian culture, slipping somewhat from the grasp of the state that had so thoroughly sought to shape it, seemed to be recombining and reconnecting with far-flung influences at a very rapid pace. Keeping up with cultural change in Tanzania has one feeling a bit like a cartoon character on a train, having to lay the tracks down out of a box just in front of the train, faster and faster in order to keep up. Dar es Salaam is the origin and the terminal point of these railway tracks.

In this chapter, I utilize Dar es Salaam's experience with the UN Sustainable Cities Program – and especially its programs for solid waste – to reflect on the confluence of neoliberalism, sustainable development, and good governance in Sub-Saharan Africa's cities at a time of alleged culture wars and clashes of civilizations. My argument is that the progressive rhetoric deployed in the Sustainable Dar es Salaam Program belies an authoritarian governance framework at its core, one whose ties to the city's colonial legacies are unmistakable. The program's short-term successes – as measured in improved rates of the collection and disposal of solid waste –came at the expense of longer-term possibilities for reconstructing the relationships between Dar residents and the local and national state. Such a reconstruction might still be possible, but a reinforced exclusionary democracy may serve the interests of neoliberal development. Exclusionary democracy in planning also simply extends the possibilities for a divisive politics of cultural difference in a city that had been relatively free of the violence of those possibilities until the last dozen years. Before laying out the case study and my argument about it, I introduce the city in the section below.

Dar es Salaam

Dar es Salaam began as a colonial imposition. It has spent much of its existence since then disregarded by its rulers. Its first plan and plat were established in the 1860s as a satellite mainland port for the Omani Sultan of Zanzibar, Majid, on top of two small Zaramo fishing villages on either side of a tidal creek. Majid's dream of making this his mainland capital were not shared by his successor, Barghash, who in effect let the place return to the Zaramo (even as late as 1957, the city was 57 per cent Zaramo), albeit with Omani customs collectors and Baluchi troops to maintain the occupation (Tripp 1997, p. 29). The German East Africa Company took control in 1887, and the German government four years later. They built a set of prominent buildings but seemed to have more interest in how the port functioned in the extraction of wealth from their plantation cash crop economy than in how the city itself developed, or how its African residents fared (Massaro 1998). British forces occupied the city in the middle of the First World War, and, upon gaining control of German East Africa under a League of Nations Mandate, made it the capital of Tanganyika Territory. But it had no legal status as a municipality, despite its size, until 1949 (Kironde 2001a). It had no representational, non-racial council until the 1961 independence of Tanganyika.

From 1961, Dar es Salaam was accorded status as a city under Tanganyika law, the first and, for most of the years since, only city in the country in legal parlance. Yet under the first (1961–64) and second (1964–69) national development plans, and under ujamaa planning (1967–85), urban development was as discouraged or disregarded as it had been under colonial rule. City plans were developed in 1968 and 1979 but largely without implementation (Armstrong 1987). A national urban development framework launched in 1980 was clearly aimed at pushing growth and investment away from Dar es Salaam, particularly with the ever-impending move of the country's capital to Dodoma (Darkoh 1994). Julius Nyerere's vision of decentralization entailed the abolition of local government, including the city council, which did not exist from 1972–1978. Hence much of the space of the city we see today – or at least where the majority of its residents live – has grown up ignored in plain view (see Figure 3.1).

From the sleepy, dusty, and rusty port I first visited in 1989, Dar es Salaam has, in fifteen years, become a constantly churning cultural engine, the heartland of *Bongo flava* Swahili hip-hop, one of the most burglarized cities on the planet, and a fiercely contested space for the politics of cultural difference. Despite the technicality of the capital being shifted to Dodoma ever so slowly, it is really in Dar es Salaam where Tanzania *takes place*, as a contestation of a national sense of culture (Lewinson 1999; Ivaska 2003; Brennan 2002). Even if a bomb crater from 1998 where the US embassy used to be has been smoothly built over, daily reminders can be seen of how edgy the city – and the country – have become, struggling with racial, religious, regional, class, and gender conflict seemingly all at once. Ubinafsishaji has spread into privateering, into the privatization of culture, into the private struggles of Dar citizens with the sheer pace and force of change and hardship (Lugalla and Kibassa 2003).

Dar es Salaam is unquestionably Tanzania's primate city, with a population estimated at 2,497,940 in the 2002 national census, growing at 4.3 per cent per year (see Table 3.1). The city officially constitutes about 7 per cent of Tanzania's population and a third of its urban population. Dar es Salaam is often listed as Africa's most rapidly growing city (Brunn and Williams 2003; Kondoro 1995). Most scholarly estimates claim that Dar es Salaam is actually growing at a rate of 8 per cent per year, and that its annual growth rate has been at or above this level for most of the last 40 years (Maira 2001, p. 36). Many analysts and policy makers contend that the census figure is significantly below the actual population, which they place well above 3 million. But this rapid growth has occurred in the absence of corresponding expansion in formal sector employment, and during a prolonged period of limited public sector investment in housing, infrastructure, or management (Kishimba and Mkenda

Table 3.1 Growth of the City of Dar es Salaam

YEAR	POPULATION
1967	356,286
1978	843,090
1988	1,360,850
2002	2,497,940

Source: Bureau of Statistics, United Republic of Tanzania

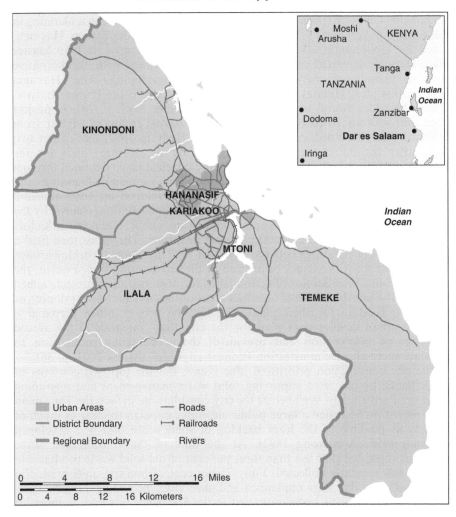

Figure 3.1 The Dar es Salaam Region
Source: University of Kansas Cartographic Services

1995; Darkoh 1994). That has left more than 70 per cent of Dar es Salaam's population to make do for themselves in terms of livelihoods and urban service provision (Lugalla 1995).

The Sustainable Dar es Salaam Program

That scenario of extensive expansion in hungry times was among the causative factors for Dar es Salaam to be chosen to serve as the global pilot city for the

Sustainable Cities Program (Halla 1994). After the director of town planning in the central government's Ministry of Local Government, Efraim Hayuma, asked for UN-Habitat assistance in 1989 for the revision of its 1979 Master Plan, Habitat suggested instead that the city experiment with an alternative strategic planning framework its offices were developing. After Hayuma rejected this idea, Habitat went around him, directly to the Prime Minister, where their idea found more accepting ears (Halla 1997, p. 28). A project document was signed in April 1991, and an external technical advisor from Habitat, Chris Radford, came in November. Early in 1992, more than two years after the initial suggestion, Habitat gave the startup funds for the Sustainable Dar es Salaam Project (SDP) and coached its city council through the five steps of its new framework, the Environmental Planning and Management (EPM) process. The SDP office developed an environmental profile of the city (UNCHS 1992), then held a city consultation, chaired by the Prime Minister, with 350 participants from private, public, and popular sectors to prioritize urban environmental planning problems. The office then held a series of mini-consultations between 1993 and 1997 that forged working groups of stakeholders for the issues prioritized in the city consultation. Fourth, the working groups then developed action plans for demonstration projects aimed at solving the prioritized issues for small areas of the city, and sought donors or alternative funding for their plans. Finally, the project office prepared a strategic urban development plan for the city that "integrate[d] the agreed strategies of intervention and provide[d] the coordinating mechanism to replicate successful Demonstration Projects citywide" (Maira 2001, p. 38).

The city consultation prioritized nine issues, but the top two concerns of participants, by far, were improving solid waste management and upgrading unserviced settlements. Well before the city consultation, in fact, the Tanzanian government orchestrated a large public cleanup campaign to clear as much of the city as possible of the huge backlog of solid waste that had developed (Yhdego 1995; Armstrong 1992). At the time of the consultation, some estimates suggested that less than three per cent of the solid waste produced in the city was regularly collected. Fully 75 per cent of housing units existed in settlement areas that were unplanned and unserviced, providing a logical basis for the second priority – obviously tied to the first, since this huge segment of the city had no solid waste service whatsoever (Kironde 1999). A variety of working groups developed out of the city consultation. They produced a scattering of action plans, but the main actions addressed these two main priorities.

The Sustainable Cities Program office in Dar es Salaam was situated initially within the Dar es Salaam City Council. That city council was a government appointed one at the time that the SDP started. It became a democratically elected body for a little while (November 1994 to June 1996) in the dawn of multi-partyism. The opposition National Convention for Reconstruction and Reform (NCCR-Mageuzi) even managed to garner one of the 55 Council seats in a June 1995 by-election for Manzese ward (Okema 1995, p. 6). The elected body was widely viewed as ineffective even by ordinary citizens, but its demise was made certain by the degree to which it became a nuisance to the central government controlled by the CCM. In mid-1996, Tanzania's Prime Minister dissolved the council and replaced it with the Dar es Salaam City Commission, whose membership was entirely appointed by his office.

The new Commission's Director, the charismatic Charles Keenja, quickly went about the process of implementing as many of the action plans as he could, according to his own vision of them. He is chiefly known for implementing the privatization of solid waste collection and disposal in the city, a process that had proceeded slowly under the SDP in its initial years before Keenja's appointment. As a result of the program's combination of private and popular sector companies in solid waste collection and the Keenja commission's privatized street-cleaning operations, Dar es Salaam became visibly cleaner between 1996 and 1999. The rate of collection and disposal of garbage increased from the estimated three per cent at the outset of the SDP to more than 40 per cent by the end of the 1990s (the official rate stood at 43 per cent as of August 2003). Under Keenja's forceful leadership, Dar es Salaam moved ahead to become "the first city in the world" to implement a Strategic Urban Development Plan (SUDP) "by applying the new EPM approach to urban planning" (Kitilla 1999, p.125). In 1998, the UN awarded the SDP a Best Practices Award for its Keenja-era Community Infrastructure Program (CIP) in the neighborhood of Hanna Nassif, which had received more than 6 million dollars in aid from the World Bank and Ireland Aid (Kironde 2001b, p. 66). In 2000, Keenja became the only Tanzanian individual among more than fifty people and more than sixty institutions all over the world named since 1989 to the UN Habitat Scroll of Honour, for his "successful leadership to make Dar es Salaam a safer and sustainable city." He shares space on the Scroll with an impressive and illustrious group – ENDA Tiers Monde Dakar, the National Slum Dwellers Federation of India, Millard Fuller (founder of Habitat for Humanity), the late South African Housing Minister Joe Slovo, and activist academics or planners like Jorge Hardoy, John Turner, Akin Mabogunje, and Otto Koenigsberger.

Keenja's City Commission was disbanded in late 1999, to make way for another attempt at democratically elected councils, while Keenja became Tanzania's Minister for Agriculture. Tanzania's parliament, in which the CCM holds 80 per cent of the seats, further transformed local governance in Dar es Salaam in 2000 by subdividing the city into three municipalities (Ilala, Kinondoni, and Temeke), each with its own elected council and each with its own Sustainable Cities Program office. The Dar es Salaam City Council, as of 2005, consisted of five councilors nominated from each of the three municipal councils. The city council building houses a supervisory and coordinating bureaucracy, including an EPM coordination unit, and a solid waste management unit, to oversee all three SCP offices in the city.

The Sustainable Cities Program in Dar es Salaam "became an important resource and information centre for environmental management issues in Tanzania and elsewhere in the world" (Maira 2001, p. 39). Tanzania was the first country in the world to attempt to replicate the SCP throughout its urban hierarchy. The Sustainable Dar es Salaam Project sent representatives out to Tanzania's other mainland municipalities and to Zanzibar, to "preach the

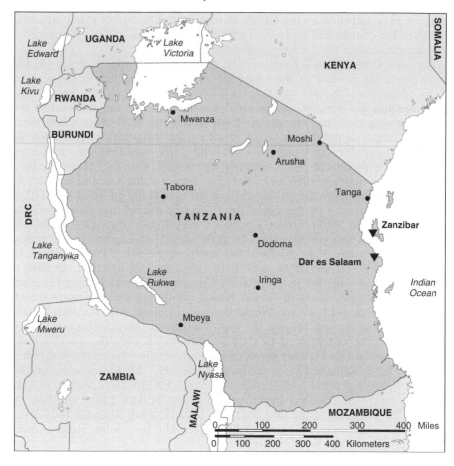

Figure 3.2   The SCP cities of Tanzania
Source: University of Kansas Cartographic Services

EPM gospel" (Kitilla 2001, p. 86). In 1998, Tanzania's Union government officially declared EPM to be the framework for urban development across the country. Sustainable Cities Program offices were established in all ten of its municipalities outside of Dar es Salaam: Arusha, Dodoma, Iringa, Mbeya, Morogoro, Moshi, Mwanza, Tabora, Tanga, and Zanzibar (Halla 1998; see Figure 3.2). The Prime Minister's Urban Authorities Support Unit (UASU), housed initially in the Dar es Salaam City Commission and then the Council, coordinated the work of all ten offices as well as the three Dar municipalities. Dar's influences across the country are clear from the fact that the director of the SDP, Martin Kitilla, was named the director of UASU, and that it not only operated out of the Dar Council building but also functioned as a de facto department within it.

From within the Dar es Salaam City Council, then the Keenja-era City Commission, and then the three new municipalities with their parent City Council bureaucracy and UASU advisors, the SCP-related offices have facilitated the privatization of solid waste collection, as well as various community infrastructure upgrading projects. The most famous programs have been those for community infrastructure in Hanna Nassif in Kinondoni Municipality and the complete overhaul of solid waste management (Kironde 2001b, pp. 60–61). Most SDP initiatives, including these, survived to 2005 only with "considerable donor assistance," and they have endured "organizational and management problems" as well as "serious clashes of personality" along the way (Kironde 2001b, p. 62). In addition to the international attention that the SDP has earned, the workings of EPM within and around it have become a subject of great debate amongst Tanzanian academics, and the overall planning discourse engendered by it has been widely discussed in the general public. In four successive sections below, I want to use the analyses and critiques of planning practitioners, academics, and citizens to develop my own assessment of the Dar es Salaam model in the context of the four story lines of this book – neoliberalism, sustainable development, good governance, and the politics of cultural difference. My assessment is based around interviews and fieldwork in May-August 2003, and analyses of documents, scholarship, and journalism related to the program.

Neoliberalism in Dar es Salaam

The city's lack of fiscal or political autonomy – and even absence of local government for it at different times – means that neoliberal policies implemented in Dar es Salaam via the Sustainable Cities Program must first be seen in the context of the broader relationship between Tanzania and the IFIs and western donors. This relationship had been generally respectful in the 1960s and early 1970s, then highly combative in the late 1970s and 1980s, and, with the acquiescence of the Tanzanian government, increasingly instrumental through the 1990s and 2000s. By instrumental, I mean that the Tanzania government gradually became an instrument of IFI and donor policies, rather than the somewhat autonomous player (albeit still with guiding hands from outside) that it had been in Tanzania's socialist era. This is a crucial, perhaps even a determining backdrop for the SDP's projects in Dar over the past decade or so.

During the period from 1961–1978, as Ibrahim Shao and his colleagues (1992, p. 2) have demonstrated, the Tanzanian economy experienced "a very favourable rate of growth" in the Gross Domestic Product, averaging more than two and a half per cent per year per capita during those years. Solid industrial growth accompanied a steady rise in life expectancy and literacy as well as substantive declines in child and infant mortality during the heyday of ujamaa policies. Shao et al. (1992, p. 3) also note that "despite the country embracing a socialist ideology in the late 60s and 70s, the WB [World Bank] was still ... guiding development policies and strategies in the country." The

IFI role in the economy is underscored by the liberalization of import controls that Tanzania carried out at the IMF's insistence in 1978 (Shivji 1991). Still, it needs to be stated plainly that for many of its proponents, "ujamaa was meant to rid the country of capitalism. It aimed, therefore, at bringing both equity and development in terms of self-reliance and self-sufficiency" (Shao et al. 1992, p. 6). Doing so required fairly massive startup investment in the infrastructure and social services (education and health care) that colonialism had denied the African population. Those are the sorts of investments that do not earn foreign exchange. The increasing influence of neoliberals in the World Bank and IMF as well as major western development agencies in the late 1970s and early 1980s showed clearly in the disdain with which much of the outsider development machinery held ujamaa, since its social programs inevitably encumbered the country with huge external debts that its very ethos objected to repaying.

The 1978 import liberalization coincided with a set of disastrous events and policies in the next few years to pull Tanzania under water. Petroleum prices skyrocketed in the world at the same moment that prices for Tanzania's agricultural exports declined. These external factors joined internal or regional problems as causative agents for the eventual imposition of structural adjustment. Tanzania's costly invasion of Uganda, ostensibly to depose its hated dictator, Idi Amin Dada, drained the treasury as the invasion turned to chaotic occupation. It is often also claimed that a "high proportion" of money borrowed for development went into "conspicuous consumption and 'white elephant' projects" that further increased the country's debts (Shao et al. 1992, p. 5).

Indeed, IFI structural adjustment was imposed in Tanzania to address only the internal factors at work in the creation of debt. From 1980 to 1986, the Tanzanian government resisted IMF and World Bank efforts to impose their policies, in effect attempting to design its own structural adjustment program. Tanzanian resistance to IFI pressures led "even sympathisers [with ujamaa] like the Nordic Countries" to suspend or cancel loans. It was only after the retirement of President Nyerere (who had engaged in regular "shouting matches" with the IMF) in 1985 that the frost began to burn off of the donor relationship again (Chege 1994). In 1986, the new President, Ali Hassan Mwinyi, signed up for the full slate of adjustments that neoliberal policy makers had demanded. These included a set of measures familiar to analysts of development around the world: policies to increase exports that would earn foreign exchange (such as devaluation of the shilling), emphasize rural agricultural development as comparative advantage, reduce inflation, eliminate government-run parastatals and government controls on the economy, and prioritize infrastructure over social services in government expenditure to facilitate trade (Ndulu and Mwega 1994). So thoroughly did Mwinyi attach himself to free market doctrines that he came to be known on the streets of Dar es Salaam as Mzee Rukhsa [Old Man Permission] for his habit of granting free rein to businesses (including his own). But "Tanzania's ownership of development policy has been dramatically reduced" since 1986, steadily "taken over by the World Bank and the

IMF" together with western donors (Wangwe et al. 1998, p. 117). Each round of rescheduling for debt payments, whether it gets called debt "relief" or not, has entailed the creation of new loans that keep the country on the hook.

Urban areas in Tanzania were dramatically impacted by the imposition of structural adjustment in the late 1980s and early 1990s, the time period in which the Sustainable Dar es Salaam Program was born (also as an imposition from outside). The impacts were chiefly negative. Tanzania had only recently reintroduced urban local governmental bodies at the time of the 1986 SAP agreement. Part of the package of reforms entailed the sharp reduction of support from the central treasury for these local bodies: they were expected to raise their budgets locally (Kulaba 1989, p. 208). Saitiel Kulaba (1989, p. 208), argued that in that context "further deterioration in the standard of municipal services – especially refuse and garbage collection, sewage disposal, mosquito control, environmental sanitation, roads, and water supply – appears inevitable." Retrenchment of government employees, when combined with the elimination of food price supports and reductions in free and universal aspects of social services meant many people suffered losses of income at the very moment when basic prices rose, out of pocket expenses for education and health jumped dramatically, and the already weak capacity of local government to supply needed services virtually evaporated.

Popular discontent did not manifest itself in riots and demonstrations in Tanzanian cities, as it did in many cities across Africa during this time. However, the exceedingly high rate at which urban parliamentarians were defeated when standing for re-election in the 1990 single-party (CCM) competitive polls, and the impromptu rallies of opposition politicians in 1991 and 1992 suggest the widespread presence of a high degree of dissatisfaction (Chege 1994, p. 54). Student activists, labor unions, and a vibrant new anti-CCM press sprang into action, leaving the CCM government in the city "unsure of whether it could ever bottle the genie it had unleashed" (Chege 1994, p. 65). Popular songs in Dar es Salaam from the period, such as Dr. Remmy Ongala's *Sauti ya Mnyonge* (Voice of the Abject Person) from 1989, echo this discontent. Ongala, from his base in the then-ramshackle alternative CBD of the city in Kariakoo, highlighted the unevenness with which the new realities were being experienced:

At school I was thrown out, I didn't have any shoes
I didn't have any notebooks.
I went to the head teacher, but he didn't listen, he just threw me out
Because of abject poverty ... in Kariakoo we know what that means, folks. ...
I didn't get to study, I have a thatch house and a goatskin bed
There are fleas underneath me, and bedbugs, too
I don't have electricity, I live by a sterno candle
I don't have a mosquito net, I am afraid of AIDS.
The powerful ones, you know them
Their entertainment is at the Kilimanjaro [Hotel]
Even now if you went to the Kilimanjaro you'd see them
Their children study in Europe (Ongala 1989; translation mine).

Dr. Remmy found himself facing deportation during the 1995 election campaign in Dar es Salaam, when he endorsed the leading opposition candidate for President, Augustine Mrema, with a hit praise song. The outcome of that 1995 election in Dar es Salaam stands in stark contrast to much of the rest of the mainland. Polls were in total chaos on election day, ostensibly for the reasons related to the poor supply lines for ballot papers. A hastily arranged re-run a few weeks later went smoothly, largely because of a de facto boycott by the opposition. Mrema had seemed certain to win much of the city had the original votes been counted, since his party had trounced CCM in the city council's Manzese ward by-election a few weeks earlier and his rallies drew tens of thousands onto the streets. The justification for the re-run seemed rather thin (Chintowa 1995). As the Tanzanian lawyer Robert Rweyemanu wrote, "CCM agreed to compete, but not to play by the rules" (Rweyemanu 1995, p. 9). With the near boycott of the Dar re-run, the new Council (save one councilor – but all soon to be replaced by the appointed Commission anyway) and the region's parliamentarians all came from CCM at a time when sentiments in the city clearly leaned the other way.

This is the setting for the startup of the most forthrightly neoliberal dimensions of the Sustainable Dar es Salaam Program. The largest of these dimensions by far has involved the privatization of solid waste collection in unserviced areas. My analysis of it below owes much to the pioneering research on it by Francos Halla, Bituro Majani, Robert Mhamba, Colman Titus, Tumsifu Nnkya, Fred Lerise, and others. The program began with a pilot project for the ten city center wards in 1994, in which a new company, Multinet, was given a contract to collect and dispose of the CBD's garbage. Multinet invested a half-million dollars of its own money in the venture, yet it still leased its trucks from the city at first, until it had enough capital to purchase its own trucks. It was, obviously, given the history of waste management in the city, completely inexperienced with its new line of work. Nonetheless, this pilot was expanded in 1996 by the Keenja Commission to 44 of the 73 wards in Dar es Salaam Region (all of the urban and periurban wards, since the other 29 were still considered rural and lacking in the infrastructure necessary for collection). The map of this 1996 expansion of privatization shows five companies and the City Council controlling all collection. Mazingira, Ltd. held six wards in Kinondoni, Kamp Enterprises had two in Kinondoni, Allysons Traders was given three wards in Ilala, and Kimangele Enterprises two wards in Temeke. Multinet continued to operate in the 10 wards of the CBD, leaving Council with 21 wards.

As of 2003, in these 44 wards, there were 21 private companies and 23 NGOs or CBOs operating to collect waste. This was down from a peak of 68 different contractors in 2000; 24 NGOs, CBOs and private companies folded in a three-year period. Although there were therefore 44 wards and 44 collecting firms, it was not the case that each firm operated in one ward. Instead, the two larger firms (Mazingira and Multinet) collected most of the waste that was being collected at all. The collection of waste continued to be "sporadic and in most cases incidental," since most contractors were "ill-equipped in terms of operating gear;" as of 2003, only these two companies had more than four tipper trucks (Majani 2000, p. 45).

In some areas, the CBOs and NGOs collaborated with these or other private companies. The community groups collected in the narrow alleys of informal settlement areas and then the private firms would take the waste out from the collection points. The solid waste management unit of the city council estimated that the higher end companies obtained 85–90 per cent of their collection fees, and the less successful ones about 70 per cent (Chinamo 2003). Others claim that "very few people or institutions were paying their refuse collection charges" (Kironde 1999, p. 85). In either case, by July 2003, an estimated 43 per cent of the waste was reaching the city dumpsite – approximately 980 tons per day.

Solid waste management under privatization has been praised by some, for instance for its apparently "efficient and effective ... guiding coalitions" (Majani 2002, p. 8). But it is highly criticized by others, not least by the main companies involved in it. The Keenja-era appointed commission receives greater critical attention. The Managing Director of the largest private firm, Mazingira, told Bituro Majani in 1999 (2000, p. 135) that "unfortunately most of the local people hate the Commission, sometimes for no apparent reason, and we find ourselves disadvantaged to reveal that we work for the Commission." Majani noted that the Keenja-era Commission did not manage "to earn the credibility" they were sometimes granted simply because "they replaced the Council (which was already under mistrust)."

Community based organizations also developed grievances with the system, as ordinary residents developed misgivings with them. Some groups, such as those in the Msasani peninsula's poorer communities that lie dotted between the richest area of the city, had their contracts stripped from them by the Keenja Commission (Majani 2000, p. 126). Other CBOs that were supposed to be responsible for neighborhood level refuse collection were instead debilitated by "misunderstandings and chaos" (Maira 2001, pp. 46–47).

The most well known among the community-based collecting contractors that survived the Keenja era is the Kinondoni Moscow Women's Development Association (Kimwoda) of Hanna Nassif ward in Kinondoni municipality. Hanna Nassif is often highlighted as the greatest success story of the Sustainable Dar es Salaam Program, for Kimwoda's work and for the Community Based Infrastructure Program there (Olofsson and Sandow 2003). Leaders of community waste collection groups formed in Zanzibar (the folks in Mkele from chapter one) and Lusaka both met with Kimwoda's leaders when the UN brought them to Dar es Salaam during the startup period of the Sustainable Cities Program in each of those cities. All of these out of town visitors shared with me their great respect for Kimwoda's women leaders, a respect also evident in the critiques of the program published by Majani (2000) and Mhamba and Titus (2001). Yet to some extent this is faint praise from a distance. The Keenja Commission actually eliminated four other community contracts in Hanna Nassif with other CBOs (Mkwoda, Sisi kwa Sisi, Jitegemee, and Kumekucha) and gave the whole ward to Kimwoda – precisely when it established itself as a chartered private company rather than a private voluntary organization. One of Hanna Nassif's key leaders told Majani (2000, p. 157) that the Commission concealed "the whole procedure from us," and that as a consequence Kimwoda met "a lot of resistance from the residents" from that day forward. Hanna Nassif's households became "objects in the solid waste collection service" in contrast to what is supposed to happen in EPM (Majani 2000, p. 164).

The appointed City Commission veered substantially away from any progressive dimensions of the EPM framework in its implementation of the privatization policy. "The implementation of strategies developed through the EPM process [was] done at the pleasure of the Commission rather than through EPM channels" (Majani 2000, p. 42). There was significant disregard for the negotiative, deliberative relationships supposed to be the bedrock of a new approach in the poor and rich communities of the city. The resultant "breakdown in communication" between ordinary residents and the government, private, and popular sectors actors had "serious implications for the social capital element in the society, that is to say that it erode[d] the norms, trust and reciprocity networks that facilitate mutually beneficial cooperation in a community" (Majani 2000, p. 121). As a consequence, a city that had at least rhetorically launched itself toward ujamaa [socialism/family-ness] became a city of ubinafsishaji [privatization/selfishness].

Sustainable Development in Dar es Salaam

The Sustainable Cities Program is often identifiable for its economic and political policy interventions, and for the ideal of EPM. But we should not forget that its impetus comes at least in one major dimension from environmentalism. The SCP purports to offer a policy toolkit that will address the Brown Agenda, a common shorthand term for urban environmental problems. At the outset of the Sustainable Dar es Salaam Program, there is no question that the city faced a daunting array of Brown Agenda crisis points. The Environmental Profile and City Consultation of the SDP did a thorough job in identifying these. At the top of the list one finds spontaneous settlements unsuitable for human occupation, near-total lack of solid waste management, invasion of ecologically sensitive lands, and a deterioration of infrastructure that then triggered environmental health problems (Sykes and Mtani 2003).

The utilization of private companies in combination with EPM-derived community-based organizations for solid waste management was seen as a central tenet of plans for addressing this cornucopia of problems. Removing the waste buildups from neighborhoods, by whatever means, would reduce environmental health risks associated with garbage in the unserviced communities. On the surface, the SDP can record some successes here. Beyond the increase in the overall rates of collection and deposition, several accomplishments are worth noting. First, the organization of a scavengers union at the city dumpsites in Vingunguti and then Mtoni, together with an intensive effort by the City Council solid waste and EPM units to forge partnerships between this union and private firms, has resulted in an impressive record of recycling. By 2003, nearly 75 per cent of glass, paper, and plastic were separated at the Mtoni dumpsite; the municipality's collection crews separate out scrap metal prior to disposal. Between 10 and 15 per cent of all collected waste is recycled via this program. Some 60 per cent of all bottles for drinking water sold in Dar es Salaam in 2003 were made from recycled plastic (Chinamo 2003; Mkumba 2003).

Second, several community-based groups have been able to keep afloat amidst competition from larger private companies, and this does represent employment for low-income residents. Agencies such as Kimwoda are required by their contract with the city to hire workers from within the communities they serve (Mhamba and Titus 2001, p. 224). The workers have received training that can, in their eyes at least, provide them with possibilities for employment when and if the particular projects they are engaged in end (ILO 1998, p. 28).

Beneath the surface, literally and figuratively, the outcomes have been "more problematic" (Olofsson and Sandow 2003, p. 61). Waste collection rates have varied substantially by neighborhood. Well to do sections of Kinondoni Municipality, such as those of the Msasani Peninsula serviced by Mazingira, had collection rates above 50 per cent. Poor communities right next to these areas in Msasani, or places like Hanna Nassif, had collection rates below 20 per cent. Private companies, logically enough, only collect from customers who pay. As noted above, the community-based solid waste groups that succeeded and endured were those that functioned effectively as private companies. Fewer and fewer of the *wanyonge* or *walalahoi* had anything to do with the new regime of waste management as time went on; less than half of Kimwoda's "confused" customer base was willing or able to pay the solid waste collection fee (Olofsson and Sandow 2003, p. 64). The visibility of areas from major roads – and proximity to them – played a key role in the Keenja cleanup, and this emphasis on visibility continues. Neighborhoods off of main roads, out of sight of donors and elites (in other words, neighborhoods where most Dar residents live), and neighborhoods near dumpsites prove to be a different story entirely.

The political relationship between dumpsite communities and the government has long been abyssmal. The Tanzanian government had an agreement and a loan for six million US dollars with Danida to build a sanitary landfill in Kunduchi, north of the city, but Kunduchi residents took them to court and prevented it (Armstrong 1992). City facilities or proposed sites at Tabata and Mbagalla were closed by the same court. Some residents impacted by landslides in Vingunguti invited the city to dump waste there to retard erosion; they were able to dump at Vingunguti for some 10 years, but were eventually stopped there too (Chinamo 2003). Wherever the dumping site moved to, the city council and the Keenja Commission have repeatedly "turned a deaf ear on" the local residents' complaints (Majani 2000, p. 49).

By 2001, the city moved to using a dump at Mtoni Kwa Kabuma, about 10 kilometers from the city center (Mkumba 2003). Mtoni dump is right on the side of a mangrove stand and tidal creek. City officials claim that they were asked to dump here by the residents right on the edge of this bluff, again because homes were eroding away. From two to three meters deep with

Figure 3.3   Dar es Salaam's Mtoni dump (2003). The truck's sign reads: Weka manispaa ya Kinondoni safi: Make Kinondoni Municipality clean.
Source: author

informally disposed waste when the city came in October 2001, the trash had piled up to 15 meters thick in a half-mile square area by July 2003 (see Figure 3.3). The dump managers were supposed to charge each vehicle a ridiculously low set fee of 1000 Tanzania shillings per ton (about a dollar). Theoretically, with 250 or so vehicles a day, the dump should be making 800,000 to one million shillings per day. What the dump was actually obtaining from 2001–2003 was insufficient for paying its workers' salaries or buying the diesel it needed to operate lifts, trucks, and bulldozers. These machines were frequently broken down, and therefore the dump workers could not actually cover and mix as they were legally expected to reduce environmental impacts.

The staff fumigated at most two times per month with hand-held pump sprayers. None of the employees had any protective equipment as of June 2003. There were several portions of the waste pile that smoldered for months. There was a huge fire that the fire brigade put out in early 2003, but with the dumping as it was, there was very little chance of putting the whole fire out completely given the combustible materials mixed together. One could see pesticides, bug sprays, aerosols, paints, and the like mixed together at the dump, as well as effluents flowing openly down into the tidal creek below. The dump was supposed to be closed in July 2003 but it was not likely to close for years.

"Really," the dump manager told me, "we haven't achieved very much" in the SDP era.

Elmalieh Elmahi (1993, p. i) found that both the groundwater and surface water at and around the earlier city dump sites in Vingunguti and Tabata were "highly polluted," with "leachate of a very high pollution strength." Mtoni, as an uncontrolled site, is likely to fare no better, with the caveat that it equally pollutes the tidal river that its leachate flows into. Rubhera Mato (2002, p. 133) showed through GIS and environmental modeling techniques that about half of the city – roughly the eastern half – has "high groundwater vulnerability" to pollution from "indiscriminate waste disposal practices." And "uncontrolled, indiscriminant dumping" remains the predominant means of disposal in the city (Ngiloi 1992, p. 75; Majani 2000). The resultant water pollution not only threatens human health directly, but also indirectly: the outflow areas from Tabata and Vingunguti dumps and almost all of the major lowland areas in the city are also productive centers for urban agriculture (Sawio 1999). Crops from these fields have exceedingly high contamination from heavy metals, and even other lowland farming zones produced crops with high toxicity as a result of water and soil contamination stemming from solid waste in surrounding neighborhoods (Sawio 1998).

It is not just in neighborhoods adjacent to the revolving set of dumps that the city tries to keep open that environmental impacts from the solid waste crisis affect human health and the quality of life. Michael Mpuya's (2000) research in Hanna Nassif, the star neighborhood of the Sustainable Dar es Salaam Program, clearly evidenced a continuation of severe environmental crises beyond the toxicity of urban agriculture crops there after the program had already been winning awards for its achievements. Mpuya focused on the lowland squatter areas of Hanna Nassif (known as Bondeni [in the lowland]) because the "laissez-faire attitude of government" in the neoliberal era had allowed for massive squatter development in hazard prone areas (Mpuya 2000, p. 6). But despite the apparent "laissez-faire attitude," under land laws that continue to this day the city government is responsible for the allocation and surveying of all plots for building, since all land is technically government owned. The government has failed miserably at this task. Wilbard Kombe's research showed that from 1978–1979 to 1991–1992, the Dar es Salaam government received 261,668 applications but only surveyed and allocated 17,751 plots, or less than seven per cent (cited in Mpuya 2000, p. 15). The rate has increased slightly since then, but Lusugga Kironde's findings from 1998 showed that only three per cent of residential land in the city had been obtained legally by the government's allocation system (cited in Mpuya 2000, p. 14). Most people in the city thus obtained and developed their plots extrajudicially.

Neoliberalism and structural adjustment are directly and indirectly culpable in increasing environmental crises in the city (Kishimba and Mkenda 1995, p. 220). For instance, neoliberalism made the construction of high quality permanent housing in these squatter areas unlikely financially. Rents for rooms in permanent structures became prohibitively expensive. The combined effect, given the high rate of growth of the city population, was "a very rapid expansion of shacks in the peri-urban areas" where "contagious diseases and hooliganism and criminality are rife" (Shao et al. 1992, p. 23). I return to Shao and company's view of hooligans and criminals in the section below on the politics of cultural difference. Their claim of increased contagious disease is more significant for my interests in this section. This is because "pests and vectors of different kinds such as fleas, ticks and mosquitoes will have direct access to people depending on the nature of housing they can afford" (Shao et al. 1992, p. 23), and unserviced neighborhoods with poor housing and uncollected wastes will have higher populations of these pests and vectors (Lugalla 1995, pp. 77–78).

Clearly, much of Hanna Nassif remains unserviced or underserviced, with large numbers of people living in poor housing on hazardous plots (Mpuya 2000, p. 7). Hanna Nassif has more than 12,000 residents, more than 60 per cent of whom live in housing areas that have not experienced upgrading or solid waste service provision. A similar percentage earns its livelihood in the informal sector. More than half of the population still lacks electricity and more than 90 per cent use a pit latrine. As of 2003, Hanna Nassif Bondeni had no tap water or electricity connection, no school, and no road to link it with the rest of the neighborhood. Hanna Nassif Bondeni, and all four major lowland squatter areas in the city, experienced severe flooding in 1990, 1995, 1997, 1998, 1999, and 2002 (Mpuya 2000, p. 61). While other areas of Kinondoni Municipality may have seen an increase in solid waste services, Hanna Nassif-Bondeni still had "considerable uncollected waste" in 2000 (Mpuya 2000, p. 91) and much of it remained in plain view in 2003. The inevitable result of this lethal combination of biannual flooding and copious stockpiles of garbage is the regular visitation of foot fungi, malaria, cholera, typhoid, and dysentery in the population (Mpuya 2000, p. 66). Cholera is now considered "endemic" to Dar es Salaam and the "burden of disease" is "overwhelming" for most residents (Lorenz and Mtasiwa 2003, p. 7). It is hard to find sustainable development underneath the rubble of this ill health.

Good Governance in Dar es Salaam

Neoliberalism expects a significant decrease in the role of the state in society, envisioning the replacement of government with governance, a broader umbrella of partnerships between stakeholders. The new era calls for a shift from "urban government to urban governance in which the involvement and participation of civil society occupies and plays an inevitable and strategic role" (Ngware 2000a, p. 7). Unsurprisingly, the crux of the debate over the SCP and EPM in Tanzania among Tanzanian scholars and planners is the role of the local and national state in neoliberal good governance partnerships. On one hand, there are those, such as Francos Halla (1998, p. 62), who argue that "attempting to institutionalize EPM by assimilating it into government machinery is not logical." Halla's point is that EPM desires the coordination of "development stakeholders through dialogue, [but] government machinery seeks to enforce laws like development control through technocracy." Therefore, to Halla (at least in 1998 – in a 1999 article co-written with Bituro

Majani and a 2003 interview with me he suggested a change of heart) "it is not possible to institutionalize EPM. EPM can only be practiced effectively if it is initiated and operated by stakeholders outside the government machinery."

On the other hand, there are other Tanzanian scholars, such as Bituro Majani (2002, p. 67), who, despite damning critiques of the SDP still see government as "the first partner among equals in the stakeholder partnerships of the EPM process." While accepting that officials and planners often retain the centralized, single-party master planning mentality of the past, Majani still sees the state as a crucial player in the development and maintenance of working groups. "Government regulates and guarantees public interest, decision-making autonomy, management, and administrative flexibility, financial autonomy, and the applicability of market forces in the whole process of accountability in the delivery of public services" (Majani 2002, p. 67).

Majani's perspective is understandable and consistent with the socialist thinking that prevailed in Tanzania for several decades. On a conceptual level, the defense of a role for the state in planning goes without saying. Yet in practical, historical terms, an EPM process that relies on substantial state involvement – as Tanzania has – runs the risk of derailing at an early stage, given popular understandings of what the state does in the realm of urban planning. Out of colonial-era and socialist-era strong-arm tactics, for many urban Tanzanians, urban planning itself is taken to mean "the expulsion of squatters and the granting of land to those who can 'develop' it in accordance with stringent building regulations" (Shivji 1998, p. 28). State involvement requires a long-term reconstruction of the relationship between the people and the state, and particularly the local state, and EPM as it has been practiced thus far only hints at such a reconstruction, from within an overarching framework that remains oppressive.

Urban local government originated in British colonies in Africa in the late 1940s as a carefully constructed means of diluting nationalist fervor with quasi-elected bodies as toothless as those of Britain itself historically (Stren 1989b, p. 22). Native administration in urban areas proved problematic for the indirect rule approaches that Britain adopted elsewhere in Tanganyika, and alternative methods stopped short of the provision of any municipal authority for direct rule. As Andrew Burton (2002, p. 99) put it, "a colonial regime unprepared for the urbanisation of the African population proved quite incapable of creating an urban administration which either satisfied its own desire to enforce the colonial will within the township, or the needs and wants of the urban African population." Local governments were racially discriminatory in their basic organizational structure, for instance in Dar es Salaam at the end of the 1940s (Kironde 2001b, p. 8). The primary roles of African agents and representatives within these bodies were revenue collection and acting as "the eyes and ears of the District Commissioner" (Molohan, cited in Burton 2002, p. 114). These extractive, spying, "weak, and poorly functioning local governments" characteristic of African cities in former British colonies inevitably came to depend on central governments for money, expertise, and legitimacy (Stren 1989b, p. 22).

Tripp (1997, p. 62) points out that on the Tanzanian mainland the "emphasis on the centrality of the state was not only a holdover from the colonial structures but also a legacy of British Fabian socialist influences, which saw the state as the great equalizer of incomes and provider of social services." Tanganyika Territory experienced a series of left-leaning governors and administrators with wide latitude to work their experiments with society into reality. But there is no question that post-colonial Tanzania's socialist order raised the bar on centralization in the 1970s and early 1980s on the mainland, ironically through Nyerere's policy of "decentralization," the centerpiece of which was the abolition of most local governments from 1972 to 1984 (Holm 1995, p. 94; Kironde 2001b, p. 9). Dar es Salaam did regain an appointed city council in 1978, and the decentralized party apparatus did enable effective social control at the local level in cities (Barkan 1994, p. 16). But Nyerere's policies of "decentralization led in Dar es Salaam to severe problems of administrative coordination and (indirectly) to extreme levels of underfunding and staff shortages with respect to most areas of infrastructure and services" for the city government (Stren 1989b, p. 23).

Local government structures have continued to change drastically through-out the period since the abandonment of Nyerere's decentralization strategies. The 1978 City Council regained full status in 1984 (Kironde and Ngware 2000). The reintroduction of multi-party politics led to a modest effort, as we have seen, at democratizing the Council, foreclosed by the Prime Minister in his June 1996 appointment of Keenja's City Commission. It is worth re-emphasizing how ironic it is that this Commission receives credit for having furthered the cause of the Environmental Planning and Management process. EPM is, by its very core nature, supposed to be a stakeholder democracy, while the Keenja-era Commission was appointed by the Prime Minister and operated in the city largely as a dictatorship. As Mhamba and Titus (2001, p. 219) put it, "unfortunately, the Prime Minister's office dissolved the Dar es Salaam City Council and appointed the City Commission without any consultation with city residents. They were thus denied the opportunity to participate in exercising power in the management of their local government." This is an opportunity Dar residents are long used to being denied.

The creation of the three elected municipal councils in 2000 for Kinondoni, Ilala, and Temeke has proven to be another false step toward decentralization or democratization. Kironde (2001a, p. 14) notes that while the division of the city into three municipalities was meant to bring government closer to the people, "accountability leaves a lot to be desired. Many urban authorities are not democratic enough in the sense that they do not operate in a transparent manner." The fiscal autonomy of these three bodies is very minimal. Between 67 and 72 per cent of all revenues consist of central government funding (Kironde 2001a, p. 39). It is not simply in fiscal terms that the dependency is obvious. As Benson Bana (1995, p. 1) put it about the larger City Council during the earlier bout with multi-partyism, the "autonomy of the council over its own employees is substantially eroded" because "the statutory instruments grant human resource management authority more to the central government than the city council."

All of the elected and appointed councilors as of 2005 were CCM representatives. All of the councils and all of the bureaucrats remained answerable to the Minister for Local Government, whose offices provided over two-thirds of their operating expenses. These moves further extend the process by which the decentralization of planning and development in Tanzania since independence has actually been a rigorous *centralization* of authority in the national state (Halfani 1997b; Ngware 2000b).

In adopting EPM as its guiding force toward a strategic development planning framework, the Tanzanian government and the Dar es Salaam City Council were supposedly opening the door to a reversal of previous planning processes. The 1949, 1968, and 1979 Master Plans for the city were top-down, elitist blueprints for "urban apartheid" that perpetuated, or at least did little or nothing to stem the "unprecedented urban sprawl" throughout their years of implementation (Banyikwa 1989, p. 83; Armstrong 1987). But despite the claim that EPM would mean a stakeholder-based approach, the SDP began – as Majani's argument above would suggest – from the assumption that "the government (central as well as local) will always be the major stakeholder in urban development" (Kitilla 1999, p. 130). Even with this firm recognition of the first-among-stakeholders role of the state, players in the government itself did not really take SDP or EPM on board. The SDP was seen as an "external project and was not fully integrated into the council" that existed before 1996 (Kitilla 1999, p. 131). Even in the Keenja era, government institutions nominated junior members as stakeholder representatives "who could not make decisions on behalf of their institutions" and "political interference from different power centers of the government" overturned the proposals of broader stakeholder working groups (Kitilla 1999, p. 131).

Some critics contend that "weak political will, overemphasis on short-term physical outputs, reluctance to share power, and the protracted nature of the EPM process" all militate against its effectiveness (Kombe 2001, p. 190). Some of these factors feed on each other. For instance, strong political will, in the person of Charles Keenja, helped reduce the protracted nature of EPM, but through an overemphasis on short-term outputs and a reluctance to share power. As Allen Armstrong (1992, p. 282) has written, "periodic 'Clean Dar' campaigns attempt to mobilise public resources and popular involvement to clean up the city. After an initial spurt of enthusiasm, official and popular enthusiasm soon wanes and any short-lived and limited impacts soon evaporate." Ultimately, most "city residents are still engulfed in filth" (Armstrong 1992, p. 282).

Charles Keenja defended the work of his appointed Commission by reminding of the situation that his Commission supposedly inherited. There was "a great shortage of funds, a weak administration that was full of workers who had no desire to work, theft, bribery, corruption, and massive bureaucracy in service delivery" (translation mine, Dar es Salaam City Commission 1999, p. 12). By contrast, Keenja claimed that by the end of the Commission in 1999, "the administration has improved, civil servants arrive on time at work, apply themselves, and usually stay for a long time at work" (Dar es Salaam City Commission 1999, p. 13).

Keenja, by this quote, was obviously pleased that his Commission had done neoliberalism the favor of making a government body operate in capitalist business mode. Yet Keenja was and is an unabashed CCM partisan, having recently slammed the entire opposition camp in Tanzania, who "because of their blindness ... live to blame and to scorn the accomplishments of the government" (*Nipashe* 2003a, p. 1). As such, it is unsurprising that opposition politicians and media outlets reserve a high degree of scorn for Keenja. Keenja's commission is said to have "left aside the work ordered for it and run for easy matters to give it a good reputation very quickly" (*Nasaha* 2000, p. 1). An article in the Islamic opposition paper *An-Nuur* (2004, p. 2) emphasized the apparent disregard with which Keenja has held his own constituency of Kinondoni since his 2000 election to parliament and appointment as minister of agriculture. "He hasn't yet come by even to debate with us what to do about our concerns," one resident claimed. Another opposition paper, *Nasaha*, cited resident Gaspar Mrutu in condemning the Commission at the close of its existence: "We were told that the City Council had failed to clean up the city, the Mayor and all of the Councilors were sacked ... Mr. Keenja and his directors were brought in. Now, they have finished their work but all of the drainage ditches are blocked." Another resident told the paper he had grown sick of the "rotten planning of the Keenja Commission," where the poor majority, still lacking an adequate system "for dumping their garbage, throws it into the drainage ditches, then the rains come and the results" are filth and disease. Asked about the glowing reputation of the Commission, one Kariakoo resident told the *Nasaha* (2000, p. 1) reporter, "maybe they are being praised for leading an era of cholera and stomach virus."

It is not simply the Keenja Commission that did not live up to – or even really attempt to match – the ideals of EPM. As has been the case throughout Tanzania (Mercer 1999 and 2003), the NGOs and CBOs the program relied upon did not prove to be the most effective purveyors of a deliberative, democratic framework for governance and planning. There is a great sense in which all of the players were "performing partnership" rather than practicing it, simply in the interests of securing donor funding, and the donors knowingly played along (Mercer 2003). The SDP's managers said that many CBOs have suffered from "a lack of transparency" (Maira 2001, p. 47) and from "conflicts" both within and between CBOs "because of vested interests among the partners or members" (Maira 2001, p. 49). The Hanna Nassif Development Association, for instance, had imploded into a set of smaller organizations (including Kimwoda) early in the Keenja era (Nguluma 1997, pp. 71–73). The years since produced continual conflict between and among the leaders and ordinary members of these organizations (Mhamba and Titus 2001, p. 225). Most prove to be "controlled by just a few people, without any established organizational framework for ensuring accountability to their constituency" (Mhamba and Titus, p. 229). Hanna Nassif remains a neighborhood where residents fear researchers, as a step toward eviction or a political threat (Mpuya 2000, p. 17; field notes, 2003). Landlords regularly harass the residents of Bondeni in particular, even for appearing to eat or dress above their station (Mpuya 2000, p. 44). Power, plots, houses, and the money from garbage collection "pass through big people's hands" only, as one resident put it (Mpuya 2000, p. 57).

According to its rhetorical commitments, the Sustainable Dar es Salaam Program has attempted to reform planning processes to move towards the modified neoliberal vision of good governance embodied in the EPM framework. It has certainly struggled to produce these reforms, and the struggles began right in its office setting. The SDP was seen in the City Council and the Keenja Commission as outside of normal council business, with an outside budget. "People would see what we were doing, councilors or people in the city administration, and they would say, 'oh, well, that's *their* project. We'll keep doing our business as usual.' They had little awareness of how the information systems or the planning approaches of the Sustainable Cities Program might aid them" (Nyitambe 2003). Decentralizing the offices to the three new municipalities, for those who genuinely believe in the EPM model as liberatory and democratic planning, frustrated matters anew because it meant starting the process of educating their fellow bureaucrats all over again (Nyitambe 2003).

Much of what remained of the SDP at the Dar es Salaam City Council in June 2003 was stuck in cubicles accessible only from a rear exterior staircase. The man who was the driving force of the office for much of its heyday, Martin Kitilla, shrugged off the downsizing and relocation his office had experienced. "You know ... in 10 years, we garnered a lot of resources but now the equipment just rots and we are here in this small space" (Kitilla 2003). Despite that adjustment, Kitilla still felt that the EPM ideal has changed minds. "People liked the TV and radio shows we did, because it wasn't big people and politicians but people from the streets who were talking, saying, 'We did this, then we did that,' and people began to sense that EPM is real; it isn't just more talking" (Kitilla 2003). Other people – here meaning councilors, companies, and elites – began to understand that they needed to consult their stakeholders, if only to cover their backs. "The councilors have a lot of power. The heads of departments [in the City Council bureaucracy] like EPM because it protects them when the councilor comes. They can say, 'it wasn't me' and point to the agreed recommendations of a working group of stakeholders: 'look, this is what we agreed to'" (Kitilla 2003).

But the overall evidence suggests that these reforms have been "rather slow particularly at the central government level, and the relationship between the state, civil society, and the private sector is still characterized by state control, subordination, exclusion, lack of transparency and accountability" (Lerise 2000, p. 88). The EPM coordinator for Dar es Salaam suggested to me that some officials – she did not have any evidence and was not naming names – did not like the EPM participatory process because it was "*too* transparent. With so many eyes watching the processes there is no way to cheat the system" (Mlambo 2003). Still, she admitted that it was still not a process "owned by or driven by communities, but rather one dependent on the drive of the Council" (Mlambo 2003). In many cases, the bottom-up ideas are shifted by donor-driven initiatives or elite priorities, too (Mtani 2003). Lerise (2000, p. 89) therefore concludes that "the new scope of urban planning is yet to be fully adopted and practiced in Tanzania." To Lerise (2000, pp. 106 and 110), "planning as a collaborative activity" has yet to supplant "planning as a monopoly of the government" or private sector elites (Lerise 2000, pp. 106 and 110).

It is not just the way the Sustainable Cities Program has been taken up within the state apparatus, but also the way it has related down to the people that has figured in its substantial limitations. On the whole, "communities are not taking keen interest and actively participating" (Kitilla 2001, p. 105). Ultimately, a bottom-up program purportedly based on decentralized and participatory concepts and processes became a top-down, hierarchical framework manifesting the "unrelenting dominance of the [central] state sector" doing the bidding of its paymasters (Halfani 1997b, p. 145). Perhaps the most damning critique is that, after more than a decade and millions of dollars of aid from a half-dozen major donors in Dar es Salaam, "very little has been achieved in terms of poverty alleviation, improved access to municipal services, and better living conditions" (Lerise and Ngware 2000, p. 124).

The Politics of Cultural Difference in Dar es Salaam

Alongside the government-directed efforts at reformulating planning through the SDP, Dar es Salaam's informalization has expanded apace: "a lot of action is undertaken spontaneously by city residents, many times contrary to the wishes or plans of public authorities" (Kironde and Ngware 2000, p. 2). Indeed, Mohamed Halfani (1997b, p. 141) has said that the "informal system of governance is so powerful that it is able to define its own distinct morphological patterns of space organization which overlaps with the official version." If Halfani has it correctly, the informalization of Dar es Salaam means that there are in effect two cities – official and unofficial – and despite the big stick of the former it is the sound of the latter walking softly that carries the process of urban development with it (Burra 2004). The escalating politics of cultural difference in Dar have made that sound much louder in the last few decades, and the SDP has generally failed to capture its spirit.

Dar es Salaam often fools its visitors into thinking of it as a city that lives up to its name (a Haven of Peace). Beneath its supposedly sleepy exterior, Dar has long been a much more contentious place. Its residents colloquially know it, after all, as *Bongo* (Brains), because it takes full command of one's wits to survive, let alone to succeed here (Sommers 2002). This is not completely a new story: Burton (2002, p. 99) writes of colonial Bongo that "major riots occurred in 1947, 1950, and 1959, and petty forms of disorder were ... common, associated with delinquency, criminal activities, inter-ethnic tensions, labour disputes, and, in the 1950s, anti-colonialism" (Burton 2002, p. 99). Criminality has expanded at a staggering rate in the last decade, though, prompting a shift in priorities within Habitat's relationship to Dar, toward its role as a model city in the Safer Cities Program (Mtani 2003). The emphasis from the Nairobi offices, then, is no longer so much on making Dar *sustainable* as it is on making it *safer*, and the research of this new program clearly suggests why. Fully 43 per

Figure 3.4   Sungusungu (neighborhood watch) cartoon: Mouse: 'Hey, I'm the neighborhood watch. You look like thugs; where are you going to commit crimes?' Thug: 'The little one is dangerous.'
Source: *Majira* (Dar es Salaam), June 20, 2003

cent of Dar residents reported falling victim to home burglary between 1998 and 2003, a stunning percentage that is nearly one and a half times the burglary rate (29 per cent) reported for Durban, in allegedly crime-ridden South Africa (Robertshaw et al. 2003). One letter to the editor lamented that the city seemed to be becoming an "abode of vagabonds and muggers" (Hussein 2003).

Majambazi (armed thugs) have become regular visitors, particularly to low-income areas since many well-to-do elites can hire armed response teams that discourage them. Government measures for confronting crime, like the Safer Cities Program, were openly scoffed at by the general public (see Figure 3.4). There was widespread belief that the police were involved in the networks that traded in stolen vehicle parts (Robertshaw et al. 2003, p. 39). Yet some of the crime is calculated by a different math, since one of the most prominent victims of majambazi in Tanzanian history was indeed Charles Keenja, who was bound and gagged in July 2003 while his office at the Ministry of Agriculture was robbed of 80 million shillings (*Guardian* 2003).

Clearly, there is a degree to which the rise of violent crime in the city is tied to neoliberal reforms, and there is a degree to which such crime is highly politicized. The combination of fractured oppositional politics and the *ubinafsishaji* of neoliberalism wreak havoc on social unity (see Figure 3.5). Paul Kaiser (1996, p. 227) analyzed how much we can claim the existence of a "causal link between liberal economic reform and social unity" – or actually, the growing lack thereof – in Tanzania, and concluded that this causal link was

Figure 3.5   **Ubinafsi (privatization) cartoon: On the cliff wall: ubinafsi: privateness, selfishness. On the bus: upinzani: The Opposition. At the chasm: Safari Njema: Have a Pleasant Trip. Road sign up ahead: Umoja: unity.**
Source: *Nipashe* (Dar es Salaam): June 30, 2003

quite evident. Ujamaa-era Tanzania had conscious policies designed to "protect the cultural integrity of a nation with over 100 ethnic groups and substantial Muslim and Christian populations of black African, Asian, Arab, and European descent" (Kaiser 1996, p. 230). Ujamaa policies became "a source of positive encounter" between Muslims and Christians, setting Tanzania apart from many other African countries rent asunder by Muslim-Christian divisions (Rasmussen 1993, p. 109). Kaiser noted that the substantial increases in income inequalities that began with structural adjustment in the mid-1980s also gave rise to anti-Muslim, anti-Asian politicians like the Rev. Christopher Mtikila in Dar es Salaam. Muslim mihadhara (public rallies) led by certain imams countered with increasingly strident anti-Christian diatribes (Mbogoni 2004, p. 171). Mtikila's followers in Kariakoo looted Asian Muslim shops in 1993 immediately after a police sweep had rounded up dozens of street peddlers (*wamachinga*) for trading without licenses (Kaiser 1996, p. 233). Muslim extremists responded by burning Christian-owned informal sector pork butcheries in the city (Mbogoni 2004, 171).

Tensions have remained high between the Christian and Muslim militants in Dar es Salaam ever since. During the 1990s, African Muslims Agency, a highly politicized and puritanical missionary movement based in Kuwait, together with other agencies of political Islam from Saudi Arabia, Sudan, Pakistan, and Iran, established a significant presence in Tanzania, and particularly in Dar (Mbogoni 2004, p. 150). Several prominent mosques, especially Mtambani and Ilala-Boma mosques, came to be associated with fiery sermons by imams such as Amiri Musa Kundecha and Ponda Issa Ponda, denouncing Christianity or US and Israeli imperialism, and equating Tanzania's government-appointed Islamic leadership with US-Israeli stooges. Islamic militancy in the city gained international attention with the August 1998 bombing of the US embassy. The plot for this bombing has been linked to the al-Qaeda network, but it also involved at least one Tanzanian national wanted for the killing of Tanzanians at the Oyster Bay site. Intensive security cooperation between the US FBI and CIA and the Tanzanian government surfaced with a vengeance after the September 11, 2001 al-Qaeda attacks in the US. At the DC Barber Shop, around the corner from the new US embassy in Msasani, barbers developed a new "FBI cut" with the increasing customer base of US agents.

Despite large street demonstrations and Tanzanian government statements opposing US military activities in Afghanistan and Iraq, this cooperation continued in the writing of a new security law for Tanzania and the rounding up of suspected terrorists in the country. The deportation of the director of Tanga's Al-Haramain Islamic Foundation of Tanzania and the arrest of Sheikh Ahmed Said Abry of Iringa in May 2003 sparked widespread protests in Dar over the new law. "If there are now those who accept being ruled by America, let them sit beside the enemy; us Muslims have chosen to defend our Tanzania," Sheikh Kundecha exclaimed (*Nipashe* 2003b, p. 1). The anger over American heavy-handedness had significant depth in the Msasani peninsula neighborhoods surrounding the new US embassy site, where residents complained that the US obtained a huge plot with few strings attached in very short order. Every imam but one refused a May 2003 peace offering from the embassy of free prayer mats for each mosque on the peninsula. Attendance at prayers at the mosque that accepted the mats dropped considerably.

The Ilala-Boma mosque and its Sheikh Ponda Issa Ponda took the next major action, occupying Ilala's Karume graveyard in August 2003, along with Muslims from other political mosques. The occupation lasted for a week, as a move to defend the graves from supposed desecration in the construction of the municipal offices for the new Ilala Municipality (Malera 2003). The court case to resolve the dispute continued into 2004, long after the occupation had ended. Although reasoned moderates in Tanzania's media suggested that "both sides must sit together and think about the whole question instead of showing each other their machismo" (*Majira* 2003, p. 6), the government and these representatives of the Muslim community seem to have irreconcilable differences, stemming from the contemporary global dynamics as much or more than from the last dozen years of religious conflict in the city. As Gontardo Matavika (2003, p. 11) has written, "in Tanzania, the process of globalisation has strained the relations among people by engendering contempt for others" in the competition "for the limited resources at hand."

A full understanding of these contemporary politics of cultural difference in the city, though, requires a deeper historical memory. Colonialism denied Africans a sense of belonging in cities, let alone rights to own a part of it. To the African urban majority population in Dar es Salaam, "density" became the language by which racial segregation was encoded, with Africans accorded the low-lying, densely populated Zone C of Kariakoo and its environs (Armstrong 1987; Banyikwa 1989; De Blij 1963). Under colonial rule in Dar es Salaam, Africans had to make "something out of the city other than what was expected" with fluidity and informal circulation through "loopholes and under-regulated spaces in changing colonial economies" (Simone 2001b, p. 20).

In Dar es Salaam, informality, fluidity and circulation became not just tools of survival but "forms of societal noncompliance with the state" after "decades of heavy state-directed top-down approaches to development" begun under colonial rule and intensified under ujamaa after 1967 (Tripp 1997, p. xv). Aili Tripp (1997, p. 1) begins her study of Dar's informal economy with a marvelous anecdote about a group of "perfect strangers" riding in an illegal mini-van transport stopped by the police, whereupon the strangers

> spontaneously transformed themselves into one big, happy family on its way to a wedding and started singing, clapping, and making shrill, ululating sounds, as is the custom for people on their way to celebrations. The police, unable to charge the driver for operating a bus on a commercial basis, had no choice but to let them go.

Such "daily evasions of state control" (Tripp 1997, p. 1) occurred and continue to occur as regular means to "make do" in Dar es Salaam. Forceful campaigns of demolition in squatter communities regularly reinforce popular anti-state convictions. Antagonism toward and noncompliance with state agents, while not universal, extend beyond the sphere of the urban economy and into those of politics and the environment. The elections in 1995 and 2000 suggested a strong CCM dominance in Dar es Salaam; the pulse of the back streets of Kinondoni, Temeke, and Ilala suggest something quite different, and in fact at times possibly something closer to precisely the opposite.

Msasani's Muslim majority, for instance – those who pray at the mosques that refused the US embassy's prayer mats – jokingly refer to their municipal council and parliamentary representatives as the ambassadors to the *wazungu* [whites], rather than as their own elected representatives. The most popular songs in the city in 2003 were Bongo flava hip-hop's mocking screeds against local government, such as Profesa Jay's *Siyo Mzee [Not so, old man]*. In this rap, the Professor plays a corrupt local politician returning to his constituents with a stump speech full of unfulfilled promises that the people, represented by the chorus in this call-and-response piece, reject. The song ends like this:

Profesa Jay: Citizens, are we together here?   *Chorus*: It's not so old man.
Tanzanians, have we understood each other there?   It's not so old man.
Dhow-pushers and voters over there?   It's not so old man.

Farmers, are we together over there?	It's not so old man.
Students, have we understood each other there?	It's not so old man.
	(Ghassany 2003a, p. 4).

This rap's huge popularity, alongside other diatribes like *Traffic Cop*, showcases the disdain Bongo's majority felt for the state, and local government in particular, in the first decade of the 21st century.

In environmental terms, the circumstances of service provision under neoliberalism leave Dar residents with little choice but to do for themselves in water supply, solid and liquid waste pollution, and environmental health (Kishimba and Mkenda 1995; Lugalla 1995). Most of the areas available for African settlement are set up to accommodate hazardous and marginal livelihood strategies. In 2003, more than half of all Dar residents "dr[a]nk contaminated water or live[d] in filthy areas" (Kasumuni 2003, p. 5). About a third had no toilets (Faya 2003).

It can hardly be said that Bongo people did nothing about these circumstances. Letters to the editor complaining about environmental conditions and environmental planning policy failures have appeared regularly in the city's Kiswahili dailies. The number of non-governmental organizations, community-based organizations, cooperatives and hometown associations seeking to cope with environmental issues from outside of any engagement with the SDP expanded in the 1990s (Byekwaso 1994; Kyessi and Sheuya 1993; Kiondo 1995). Over the course of a decade, for example, an environmental CBO in the peri-urban settlement of Makongo – Majudea (Makongo Juu Development Association) – was able to link with the urban planning department of the University College of Lands and Architectural Studies to defend its tenure and replan its own land uses (Burra 2004; Lerise 2003). In the operations of non-state actors like Majudea, we see both their flexibility in forming and maintaining local alliances and their increasing engagements with extra-local entities, even international organizations and internet linkages. Yet instead of being a true vision of an alternative city, really, most of the non-state urban organizational energy has exhibited a "narrow range of effectiveness" that serves "merely as palliative measures to reduce the poverty sting" (Halfani 1997b, p. 154). If all of Dar es Salaam's neighborhoods had a Majudea, and all the Majudea-like groups worked together to plan the city, a genuinely progressive, decentralized, and deliberative planning system would be in place. But realities in Dar es Salaam are far from that vision.

By the end of the 1990s, "a sense of deep alienation prevailed in urban centers" in Tanzania, where alternative, local level community based participatory development strategies had mostly "degenerated into organizing for survival" (Halfani 1997b, p. 123). In a context that entailed a complex and changing combination of "hostile, repressive, adversarial, extractive, competi-tive ... and indifferent" state-society relations (Aina 1997, p. 427), it is hard to imagine broad popular engagement with a "formalized and technocratic" (Aina 1997, p. 421) urban management strategy like the SCP in Dar.

It would be almost ridiculous to claim that the SDP caused an escalation of the politics of cultural difference. Yet its analysts – critics and supporters alike

– have as yet paid little heed to the cultural-political – or indeed historical – context of its implementation. Donors and the planners who have implemented it give short shrift to the intense arena of religious, regional, or identity politics that the city has become. So much is said to depend on popular sector NGOs without any note that some three-quarters of all registered community groups in poor neighborhoods of Dar es Salaam are religiously affiliated, most with mosques hostile to the government (Kiondo 1994). Likewise, critics and supporters note only the history of post-colonial governance failures, without the deeper connectivity to the colonial era and the broader problematic that this deeper look presents. Dar residents "have long practice in how to conduct their own affairs. A disruption of this 'cultural' pursuance" is implicit within the SDP's framework of operation (Majani 2000, p. 37).

Conclusion

Dar es Salaam has served as a model city for the world in the United Nations Sustainable Cities Program. It effectively models a number of themes for this book, as well. The first of these is that the neoliberal context of the SCP's implementation shapes its policy outcomes toward a highly uneven distribution of participation and of development benefits. Second, the sustainable development rhetoric attendant with the Sustainable Dar es Salaam project, when viewed from a livelihoods-oriented urban political ecology perspective, produces only limited pockets of environmental improvements, and the participatory dimensions are severely constrained by the neoliberal aspects of the relevant policies. Thirdly, the "global governance" agenda to which the SDP is linked is still "a movement … which remains incomplete and often incapable of application," as the economic historian Bill Freund (2001, p. 724) has written. Dar es Salaam remains a very long way away from a deep democracy that builds consensus from below.

With EPM now central to the teaching curriculum of the University College of Lands and Architectural Studies that trains almost all of Tanzania's urban planners, "a new perception of the role of stakeholders has been established" across the country (Sykes and Mtani 2003, p. 6). The framework certainly helps planners used to a dominative central government controlled system to recognize that "there is no planning without conflict" (Lerise 2003). But, ultimately, as Fred Lerise put it in a conversation with me, "we are finding that participatory planning does not fare well or go far when you have changed a few minds at the mid-level in government. The top still has not changed." It may be, indeed, that the bottom – meaning the people – has not changed much either, in terms of how they see the very marginal possibilities for their own involvement in planning their communities.

In Dar es Salaam, though, the "floodgates" are already open on political change (Neumann 1995). The drive for political liberalization has created some space for the defense or negotiation of peoples' resource rights, and for the reconfiguration of environmental management in the city, as the Makongo Juu example suggests. The democratization of Tanzanian society, even if it is reflected very poorly in the multi-party politics at the city level in Dar es Salaam, has still meant that "the state is no longer impervious to forces from below" (Freund 2001, p. 736). One chief manifestation of this appears in the highly contested and politicized character of cultural differences among city residents. As much of the political vehemence is directed at both the national and municipal government, it is truly hard to fathom an EPM process succeeding that depends on popular trust in state institutions. If this can be said of the model city, it is even more the case in its sister city across the Zanzibar channel, to which I now turn.

Chapter 4

The Mirror that Zanzibar Holds up for the World

I want an open and transparent conversation. We are now the regional headquarters for East Africa as a World Heritage City – we need to set a clean example.

(Keis 2003)

Introduction

With more than three million US dollars from the UNDP and UN-Habitat alone, Tanzania's Urban Authorities Support Unit (UASU) has led the effort to extend the Environmental Planning and Management (EPM) framework from its pilot program in Dar es Salaam out to all of Tanzania's municipalities. Its longest and most arduous efforts in this extension have been directed at Zanzibar city, some sixty kilometers northeast of Dar es Salaam in the Indian Ocean.

UASU's initial efforts, in Zanzibar and the other cities, apparently did "not go down well with other sister municipalities" that resented its intrusion, "bearing in mind that they were not answerable to it" (Kitilla 2001, p. 87). Later programs in a half-dozen mainland cities have received funding from the Danish International Development Agency (Danida), and the money clearly raised the profile of the solid waste management programs directed by the Sustainable Cities offices in these cities, particularly Arusha, Moshi, Iringa, and Mwanza. This aid has not flowed to Zanzibar. Not only has Danida not funded any similar project in the city, but a whole host of other donors have shied away from supporting the Zanzibar Sustainable Program. Even the United Nations and German aid moneys that have kept the program afloat have come and gone in comparatively small amounts. The total aid over time to the whole program is less than 500,000 US dollars – less than one-tenth of the aid to the Hanna Nassif Community Infrastructure Program in Dar from Ireland Aid alone. "In Zanzibar," UASU Director Martin Kitilla sighed, "the political will is just not there."

In examining the Zanzibari case in this chapter, I suggest that the problems have gone well beyond the issue of political will. I analyze the case in the context of neoliberalism, sustainable development, and good governance rhetoric, as well as a time of very bitter politics of cultural difference. Rather than emerging as a tool for creating a more inclusive city with deliberative planning processes, in the hands of the Zanzibar Sustainable Program the EPM framework has been at best an additional ball introduced onto a soccer

71

pitch in mid-game: a goal might be scored here and there, while ignoring the bloody and unjust match being played on the rest of the field with the other, weightier ball.

Before I turn to the deeper analysis, I first introduce Zanzibar city and the Zanzibar Sustainable Program in the section below. Zanzibar's unique stature, given its name redolent with many imaginary representations it only thinly deserves, requires a certain amount of demystification and normalization, which I seek to do here, even while articulating why this is still an important case to examine.

It all Comes Together in Zanzibar

Fifteen years ago, one of my professors, Ed Soja (1989), published a groundbreaking book, *Postmodern Geographies*, with a now widely known chapter entitled, "It all comes together in Los Angeles." Through the influences of a range of scholars, including Soja and his students, a great deal of urban research (if not, of course, exactly *all* of it) has come together in Los Angeles. By the close of the century, Los Angeles achieved a status to rival that of Chicago in an earlier era of urban studies, as a laboratory for assessing processes of urban development and social change concentrated there but emblematic of trends across cities in the whole world (Dear 2001).

Zanzibar, with about 400,000 people including its suburbs, is a far less significant city in the world of ideas or the world of commerce, in comparison to Los Angeles. It is unlikely that we will ever study a "Zanzibar School" of urban theory. Zanzibar city resonates with shallow echoes of a history greater than its present, just as it reverberates with romance and exoticism through countless travelogues everywhere. It is also actually a fairly ordinary African city. In fact to me it is the perfect ordinary city to serve as a laboratory for examining our extraordinary times, for Zanzibar is a microcosm of so many of the world's most pressing concerns and most dynamic trends. Since the mid-1980s, neoliberalism, sustainable development and good governance have rocketed into a place once claimed as "Africa's First Cuba" (Kharusi 1967). The legacies of and backlashes against the authoritarian tendencies of that "Cuban" time have hurtled forward into a feverish age of politics based around cultural difference, in a city that is now five times its size at independence. Africa's first Cuba starts to look more like Africa's next Jamaica, with a huge push toward luxury tourism – but all at the very moment when Islamism, regionalism, homophobia, anti-westernism, racialism, and gender conflict surge ahead with a vengeance. Name the form and shape of the skirmishes in the culture wars of the world in my lifetime, and Zanzibar has played host to them. And, importantly, it has played host to what seems to me a very typical attempt, in the current global context, to implement the Sustainable Cities Program's ideas and frameworks.

The polity of Zanzibar consists of two main islands, Unguja and Pemba, with a combined population of a little more than a million people, about 40 per cent of whom reside in the vicinity of the city of Zanzibar (see Figure 4.1).

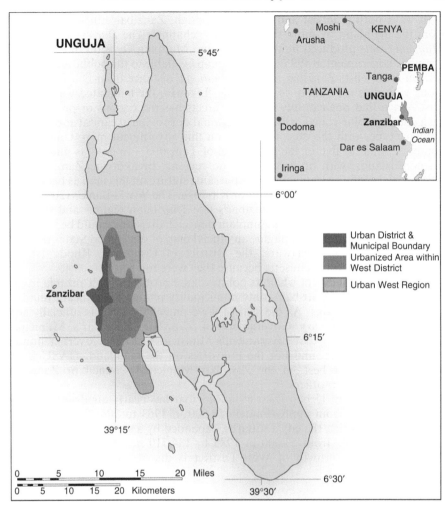

Figure 4.1   Zanzibar City and Unguja Island
   Source: University of Kansas Cartographic Services

Since April 1964, Zanzibar has been a semi-autonomous partner with the mainland (the former Tanganyika) in the United Republic of Tanzania. This United Republic is a state that in fact warily links two distinct nations. Within the Union constitution, Zanzibar retains its own presidency and government ministries, elects its own house of representatives, and sends a delegation to the Tanzanian parliament that is proportionally much larger (nearly 20 per cent) than its relative population (three per cent) would warrant. Its government receives a 4.5 per cent share of the revenues gathered by the Tanzania Revenue Authority, again slightly more, percentage-wise, than its population share.

In ways both practical and symbolic, though, Zanzibar still seems divided from the mainland by more than a narrow channel. The *Dar es Salaam City Map and Guide* produced by the Surveys and Mapping Division of the United Republic government in 1995 – the main map available to the public for that city even in 2003 – lists Zanzibar as an *International* distance from Dar es Salaam, along with Aden, London, or Singapore. Likewise the laminated wall map of East Africa that one can buy on Dar's streets has the regions or provinces of each country color-coded, with every regional or provincial capital appearing in boldface type, with the exception that, on this government map, Unguja and Pemba – home to five of the 25 regions of Tanzania – appear as unaffiliated offshore islands, with no color code, no regional breakdown. Zanzibar city receives no boldface treatment; it has the same stature on this map as backwater mainland district towns like Ifakara or Kasanga. The World Bank's (2001) map of Tanzania likewise ignores Zanzibar's five Tanzanian regions and regional capitals. Lest the cartographers complain that such divisions would be too small for the scale of these maps, imagine an official map of the United States that did not bother to show Delaware, the District of Columbia, Rhode Island, Connecticut, and Massachusetts because they were too small.

Thusly forgotten except when its politics cause problems for Tanzania in the international arena, Zanzibar's only municipality retains a significant number of capital city functions. As the seat of government for what is officially still called the Revolutionary Government of Zanzibar, Zanzibar's Municipal Council is overseen by that government's Ministries of Local Government and of Good Governance, and not the ministries on the mainland. UASU is an absent presence, at best, in the Zanzibar Municipal Council or Zanzibar Sustainable Program offices.

The population of Urban District in Zanzibar has nearly quadrupled in the post-colonial era, from approximately 58,000 in 1963 to 206,292 in the 2002 census. West District, the city's suburbs, expanded by a similar amount, but in just fourteen years, from 53,000 in 1988 to 184,710 in 2002 (see Table 4.1). Almost all of the growth of West District is due to an expansion of the

Table 4.1   **The growth of Zanzibar City**

YEAR	POPULATION
1931	45,276
1958*	69,024
1978*	142,041
1988*	208,137
2002*	391,002

Sources: Bureau of Statistics, United Republic of Tanzania; Census data for the Zanzibar Protectorate, Zanzibar National Archives
* Figure includes population of both Urban and West Districts.

urbanized area of the city. The West District wards near the boundary with Urban District are completely a part of the fabric of the city in every way but this district line: these are the places whose populations have exploded since 1988, wherein almost all of the population of West District is to be found. Urban-West Region, the entity that combines the two districts, essentially functions as a unified city of about 400,000 people – except for the fact that the Zanzibar Municipal Council has no legal right to operate in West District (Myers and Muhajir 1997; Myers 1999; ZSP 1998). The overwhelming majority of the new residents to arrive in Zanzibar city since the 1980s, indeed 45 per cent of the urban population, now live in under-serviced and unplanned neighborhoods on the fringe of town beyond the municipal boundary (ZSP 1998). It is this context of unprecedented growth and sprawl, and the overlaps and mismatches of administrative authorities to deal with them, that led to the creation of the Zanzibar Sustainable Program (Figure 4.2).

The Zanzibar Sustainable Program

Zanzibar's Sustainable Cities Program was founded in essence as an inter-agency think-tank between Zanzibar's central (Revolutionary Government of Zanzibar) and local (Zanzibar Municipal Council) governments, led by the somewhat autonomous and intellectually-minded Zanzibar Commission for Lands and Environment, or COLE (Myers 1996; Myers and Muhajir 1997). The ZSP grew out of the efforts of the COLE's Department of Urban Planning and Surveys to garner funds for localizing Habitat's Agenda 21. The Zanzibar government actually sent COLE staff to Dar es Salaam to learn about the SCP operations as early as 1994, and COLE utilized aspects of the EPM approach through 1994–7, well before Zanzibar formally signed the UN's SCP document in March 1997. The UN only began funding the city as an SCP participant in early 1998. Six years later, it was hanging on to that status, somehow.

From 1989 until its dissolution in early 2001, COLE was the leading force within Zanzibar's SCP and the implementation of an EPM-influenced planning framework, and COLE seconded key personnel to run the project. In January 2001, though, Zanzibar's President disbanded the Commission, moving its once-powerful Environment Department to the Ministry of Agriculture and the Lands and Urban Planning Departments to the Ministry of Water, Energy, and Construction. This action threw much of the burden for attempting to continue the ZSP onto the Zanzibar Municipal Council, where the UN had wanted it all along. But it took almost three more years, until July 2003, for the Council to officially be granted the mandate to run the Sustainable Cities office as one of its own offices (SMZ 2003).

The city's 1998 environmental profile, two proposition papers prepared for its December 1998 city consultation, and that consultation itself identified and prioritized a similar set of issues to those focussed on in Dar (ZSP 1998; Toufiq 1998; Muhajir et al. 1998). The ZSP was charged by the 200 or so stakeholders at the consultation with finding innovative strategies for administrative restructuring and for coping with the environmental fallout from sprawling growth in the form of solid waste management, water supply, sanitation, and

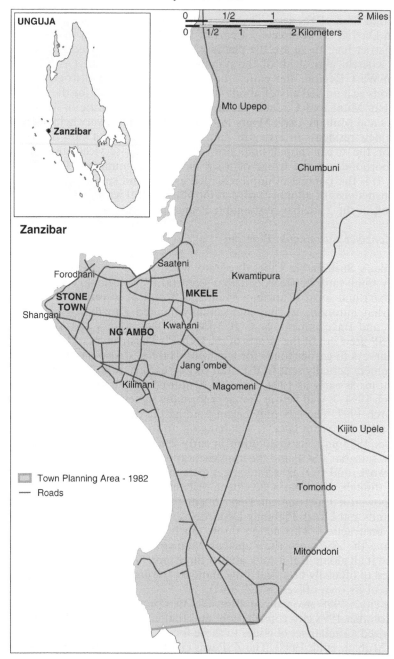

Figure 4.2   Zanzibar City, town planning area
            Source: University of Kansas Cartographic Services

haphazard construction. The ZSP declared its chief goal to be to "improve urban planning and management in Zanzibar through better coordination among different organizations and participation of all major stakeholders" in the urban environment, and it set about doing so through six working groups (ZSP 1998, p. i). One of these was quickly folded into a second one, leaving five groups. These working groups addressed solid waste management, haphazard construction, water supply and drainage, the municipal boundary issue, and municipal revenues, respectively.

In contrast with the millions of dollars in aid that have flowed into the Dar es Salaam pilot, Zanzibar's Sustainable Cities office has suffered from a rather circumscribed funding base. The United Nations Development Program (UNDP) funded the ZSP for two years, including the provision of an expatriate (Dutch) project coordinator, Paul Shuttenbelt, who was instrumental in producing the city's effective environmental profile. Shuttenbelt's "premature departure" a few months after the city consultation left the program with a Zanzibari project coordinator, Ghalib Omar Awadh, who oversaw the completion of UNDP's modest aid package in December 2000 (Nnkya et al. 2000, p. 4). It took more than a year to locate and secure a new sponsor, the German Development Service (DED), a time during which the project "nearly died" (Juma 2003). The local project coordinator was replaced by the ever-enthusiastic Sheha Mjaja Juma, who remains in this position as of 2005, but with relatively little financial capacity. The DED has supplied three Urban Management Advisors (UMA) in succession to the ZSP since 1998, with a year-long gap between the first and the second UMA, but only small amounts of money beyond this – for a few pilot projects and half of the costs of keeping the office afloat. The Revolutionary Government has provided very little funding of its own to the program.

The working groups established in December 1998 included representatives from various levels of government – Revolutionary Government ministerial officials, municipal officials, city councilors (*madiwani*), ward-level government representatives (*masheha*), and a scattering of NGO or CBO representatives. Most of the five groups met only sporadically from their December 1998 beginnings through August 2003, with some appearing more functional than others. The Steering Committee for the whole ZSP managed to meet four times in the first five years of the program. The most active of the working groups, the solid waste management group, met only a few more times than this.

Despite a relatively dormant set of working groups, the ZSP office, through Juma, the German UMAs, and a few community-based organizations and non-governmental organizations, did manage to create a few action plans. A US $6,000 project in Mitoondoni, a haphazard neighborhood in West District, created a network of standpipes for community water supply. The community group in Mitoondoni then set about finding funding for a drainage program, similar to small ones ZSP had accomplished in nearby Jang'ombe and, with the Baja Social Group, in perennially flooded Kwahani (Nnkya et al. 2000; Muhajir et al. 1998). The ZSP office itself, in the process of preparing the environmental profile, became essentially the only center of GIS mapping on the island, and for a time even earned praise from the UN for its

Environmental Management Information System (UNCHS 2000b, p. 35). For some of its personnel, ZSP became the conduit for "better coordination among different organizations and ... stakeholders" that its organizers had hoped it would be. For other active participants, ZSP has given them an opportunity to both research and in a small way act on environmental issues that their previous postings undervalued.

I have watched, and to some extent participated in, this whole evolution of events during eight research trips to Zanzibar since 1991. I have shared the bemusement and frustration with which those closest to the process have contended for a decade. I have seen the moments of enthusiasm and achievement that have come as well. By far the clearest and most comprehensive example of the coordination and capitalization on an opportunity has come in the pilot program for solid waste management in Mkele, a small, semi-planned working class neighborhood fairly close to the geographical center of the city. The Mkele case worked a little bit like the EPM process is supposed to. Mkele's city councilor (*diwani*) and leader of its only NGO, the Mkele Ward Development Committee, Kassim Juma Omar, approached the ZSP to see if they might work together to find a solution to the solid waste crisis in the ward. Through training, advice, and counsel that the ZSP provided – including a site visit for the Development Committee's leaders to Hanna Nassif in Dar es Salaam, and collaboration with the NGO Resource Center in Zanzibar – Kassim and his NGO emerged with a strategy for sensitizing and relating to the community about the solid waste issue. As I discuss in several separate entries below, the Mkele pilot is hardly a flawless model, but at least it has had substantive results beyond much else that the ZSP has done.

The ZSP has had a somewhat freer hand to consolidate the findings of a variety of rival government institutions, since it takes personnel from each of them. As a consequence, the ZSP was able to compile a thorough environmental profile of Zanzibar Municipality in December 1998 that hit at issues head on. As their first line of analysis, the authors of this profile – Shuttenbelt, Awadh, Juma, and Khadija M. Saad – critiqued the overlapping and redundant institutional responsibilities for the municipality, saying that

> the activities of individuals and institutions often reflect personal or sectoral interests rather than coordinated or collective development, ... [and] organizations are reluctant to relinquish their autonomy. ... There is no systematic manner of identifying urban development issues in Zanzibar. Policies are often absent or not followed and for this reason priority issues are often identified on an ad hoc basis. Whenever the issue becomes a nuisance, or there is a crisis or some hazard, the problem is solved on the directive of an important decision-maker (President, Minister, Director, Mayor). However, the communities at the grassroots level know their priorities quite well, and these are to improve their direct living environment.
> (ZSP 1998, p. 77)

Therefore, the ZSP has at least been capable of stating openly in print some obvious things that needed to be said but had dragged around silently in the city for decades. Moving from accurately documenting both the state of the environment in the city and cutting to the quick on the flaws in the institutional landscape established for dealing with that state, to actually planning means for addressing the environmental and political-economic problems in a participatory and democratic manner by which the "grassroots" would "improve their direct living conditions", however, has proven an entirely different matter. Attempts to replicate the Mitoondoni water project in Chumbuni, another West District informal settlement, have gone nowhere. Larger and more ambitious programs for revamping the Forodhani gardens on the main waterfront in the Stone Town area of the city and for improvement of the Darajani market area have been stifled by a lack of donor funding and intense political opposition. Even the pilot for solid waste management in Mkele has moved forward only in fits and starts. Time and again in the past decade, both with the ZSP and with the COLE's EPM-based programs that preceded it, any genuine intentions of planners for implementing a participatory EPM process from below have been thwarted from above, and by a long history of disaffection between the urban majority and the local state. The problems stem from four overlapping contextual dynamics: the rhetoric, policies, and histories of neoliberalism, sustainable development, and good governance, and the politics of cultural difference.

Before addressing each in turn, there is one absolutely vital local contextual element that warrants extended discussion. The political standoff between the ruling party (CCM's Zanzibar branch) and the Pemba-islander dominated opposition, the Civic United Front (CUF), has persisted through two bitterly contested and disputed elections in 1995 and 2000. I discuss this impasse again below in each of the sections, but I detail it here first, because it is the defining feature of the entire context of the implementation of the Sustainable Cities Program in Zanzibar.

Its contemporary outlines are as follows – recognizing for now that its roots are much deeper. As soon as Tanzania declared its intention to introduce a multi-party political system in 1992, it was clear that the islands would have a very strong regional opposition party. It was equally clear that CCM would meet that opposition with force, and the three year build-up to the first multi-party elections was a time of escalating tension and occasional violence. In the October 1995 elections, CCM won only 26 of the 50 seats in the House of Representatives, and Zanzibar's CCM President Salmin Amour retained the presidency, defeating CUF's Seif Shariff Hamad by less than one-half of one per cent of the vote. CUF won every constituency (21) on Pemba by huge margins, together with three on Unguja (two in the Stone Town area of the city). More votes were declared spoilt than the total votes that separated Amour and Hamad, and allegations of irregularities – most of them quite credible – persist to this day.

The CUF refused to recognize the results of the 1995 presidential polls. While its Tanzanian parliamentarians – at this time, all from the Zanzibar islands – conducted their affairs without interruption (one was, for a while, official leader of the opposition in parliament), its representatives in the Zanzibar House carried out various acts of civil disobedience. For instance, they maintained their attendance at roll call for one session out of every three in order to abide by attendance rules, but then left immediately after the roll was called. In 1998, they were dismissed by the CCM; seventeen CUF members, including four House representatives, were charged with treason. The Commonwealth Secretary, Emeka Anyaoku, and his assistant, Moses Anafu, negotiated a peace agreement, called the *mwafaka* (pact) between the disputants, signed on June 9, 1999, that provided a brief interlude of cool acceptance of differences. All but the four still-imprisoned CUF House members returned to regular attendance and participated in normal House business. A *tarab* (Swahili coastal music) piece commemorating that agreement became a huge hit: "now we have agreed with one another, let us shake hands, we must be glad, let us sit together and bring development," the chorus rang out. Amour and Hamad often made joint appearances in the latter half of 1999 and the first part of 2000, and on the surface it looked like the *mwafaka* would hold if the government would simply abandon the baseless treason charges.

But the new election season in the latter half of 2000 recreated the old tensions, in new ways, even as the treason charges endured. Amour's yes-men attempted to change the constitution to allow him a third term, and when that failed, the Zanzibar branch of CCM chose his loyal Chief Minister, Mohammed Gharib Bilal, as its candidate for Zanzibar President. CCM's National Executive Committee, dominated by the mainland, bitterly rejected Bilal, and imposed its choice, Amani Karume (son of revolutionary Zanzibar's first President), on the island wing of the ruling party. His family name and long loyalty to the revolutionary government, to say nothing of his low-key nature, won Karume support from many island CCM stalwarts. But given how disaffected his selection – and the manner of it – made many of Amour's supporters feel, a divided CCM seemed headed for a loss in October 2000, even as confrontations between CCM and CUF supporters escalated again. *Mwafaka* became guttural slang in the city, and *fitina* (discord) became much more the defining term for public political interaction (Myers 2002a).

That October, as the votes were being counted, apparently showing CUF victories in what were considered urban working class strongholds of CCM, the Zanzibar Electoral Commission declared the city wards to be in chaos and called for a re-run in them a week later. Even though the CUF again won almost all of Pemba (sixteen seats) by huge margins, its leaders called for a boycott of the re-run in the city. This handed CCM a much larger majority in the House and handed Karume a much more decisive margin of victory (albeit with a bigger asterisk) over Hamad than Amour had enjoyed. The CUF again refused recognition to the CCM regime, and its representatives were refused entry into the House. Then, in January 2001, the Tanzanian police confronted CUF protest rallies in Pemba and in the city before they had even begun, with brute force, saying the rally was conducted without a police permit. The police attacked indiscriminately, and a few of the protesters fought back – even to the point of beheading a policeman in Pemba, which apparently caused the government forces to engage in even greater atrocities (Arnold et al. 2002). The incidents that day constitute the worst political violence experienced in Tanzania since the Zanzibar Revolution, with at least 35 protesters dead and hundreds more seeking refuge in Kenya. The events shocked the entire republic and the international donors. Vigorous negotiations eventually led to a second peace agreement between the parties in 2002, the release of the original detainees, and the return of CUF to the House, but all of this failed to dislodge anything but the crust of the bitterness that resides on both sides. It is hard to fathom anything that has happened in Zanzibar in the last two decades without first recognizing that bitterness hanging in the air.

Neoliberalism in Zanzibar

At the same time, like Tanzania as a whole, Zanzibar has weathered nearly twenty years of economic restructuring after decades of state socialism and neglectful colonialism. The changes on the islands have come in a more concentrated and hyper-real fashion, though. The regime that seized power in the bloody coup of January 1964 in the islands had very publicly pronounced its chief goal as the eradication of the enormous class-based inequalities of the colonial era (Sheriff and Ferguson 1991; ASP 1974). Even after the April 1964 unification of Zanzibar with Tanganyika to form Tanzania, the islands were under the sway of a much more centralized and authoritarian regime than that of Nyerere on the mainland. Revolutionary Zanzibar's first President, Abeid Amani Karume (1964–1972), proudly declared that the Arusha Declaration of 1967, through which Nyerere had declared ujamaa as the policy of the land, stopped at Chumbe island's lighthouse that marks the entrance into Zanzibar's harbor from Dar es Salaam. Karume's considerable autonomy came with a significant imprint of Eastern European socialism that outlasted his 1972 assassination. East Germany had their diplomatic mission to Tanzania in Zanzibar; West Germany stayed in Dar es Salaam. The presidency of Aboud Jumbe (1972–1984) brought some clipping of the island government's wings, but Jumbe still had wide latitude to develop an extensive policy agenda of his own with his cabinet of ministers until 1977.

Economically, this autonomy meant that the revolutionary government of 1964–77 could utilize the revenue from its monopoly over the export of cloves, Zanzibar's primary export. Control over the clove account, combined with good years for clove prices and harvests, meant that the island regime was relatively flush in its first decade (Chachage 2000). Although the US retained a tiny consular office until 1978 and Britain held onto a residence that served as a de facto consulate, Zanzibar received next to nothing in direct financial assistance from the West through the late 1970s. Much of its aid had come in kind or as direct aid (rather than loans) from China, the Soviet Union, and East Germany. Consequently, when Tanzania was forced into the structural adjustment straitjacket in the early 1980s, had Zanzibar's government headed an independent regime, it probably would have had little cause to be placed in the same bind as its Union partner.

Sweeping political, economic, and cultural transformation came after the 1977 unification of mainland and island parties to form the CCM. The party unification coincided with poor harvests and plummeting prices for cloves, as well as swift constitutional reforms that included a vast expansion of the "Union Matters" (responsibilities of the United Republic government) in the Tanzanian constitution, at the expense of Zanzibar's own government (Chachage 2000). A new House of Representatives (democratically elected from amongst CCM members) produced a set of rather stunning changes of policy direction over the next fifteen years in particular, in conjunction with the Presidents who followed Aboud Jumbe. Ali Hassan Mwinyi (1984–85, before becoming the United Republic president), Adris Abdul Wakil (1985–90), and Salmin Amour (1990–2000), as well as the Chief Minister of the first few Wakil years, none other than (the future CUF's) Seif Shariff Hamad: each played a role in reshaping Zanzibar. A new Trade Liberalization Policy (1985), a Private Investment Act (1986), new, more liberal land laws (1989 and 1992), a new Commission for Tourism (1992), and a new Investment Promotion Agency (1992) signaled an aggressive opening of the country's economy. Western donors – particularly The Netherlands, Finland, (the eventually reunited) Germany, Sweden, and, to a lesser extent, Britain – began to funnel aid projects directly to the Revolutionary Government of Zanzibar, for the first time. The number of tourist visitors climbed from a mere handful in the early 1980s (less than 9000 in 1984 for instance) to more than 100,000 annually by the early 2000s (Gossling 2002, p. 4). The number of private hotels and guesthouses jumped from zero in 1980 to nearly two hundred by 2002, more than half owned by foreigners (Gossling 2002).

Seemingly overnight, Zanzibar was transformed from a socialist backwater to the hip and trendy tourist destination of choice for the elite of the world, boosted by visitors like Bill Gates of Microsoft or the beautiful people of Hollywood. More gradually, and probably more seriously, Zanzibar's fiscal autonomy withered away. Structural adjustment in Tanzania increasingly became tied to the shrinkage of differentiation between the Union and island governments, and the consolidation of revenue collection. Banking explicitly on the possibilities of creating a kind of Singapore, Hong Kong, or Dubai for East Africa, the Zanzibar government made various attempts to establish an official free port, particularly in the Amour regime of the 1990s. Although these did not come to fruition, the city became the port of entry of choice into Tanzania for many Middle Eastern importers because of the lengths to which port officials went to circumvent Tanzanian customs regulations. This "smuggling" caused the port to come vigorously to life as a somewhat piratic capitalist haven, after two decades of minimal private cargo hauls. But the IFIs decried the new "Zanzibar route" as perhaps the quintessential example of the negative effects of corruption on Tanzanian government revenue collection, and eventually the route was stifled, rather forcefully.

To the extent that the government profited from the small construction boom, the land rush, the short-lived import-export game, or the tourist explosion that accompanied the neoliberal era in Zanzibar, benefits accrued mainly – albeit of course not exclusively – in two ways. Either individual government officials or other elites profited illicitly from their connections, or the revenues flowed to the (mainland's) Tanzania Revenue Authority and Bank of Tanzania (Myers 1999 and 2002b). The latter flow came to predominate by the end of the 20th century, with exceptions here and there, like the relatively wealthy Zanzibar Commission for Tourism. In any case, though, little money remained for the Zanzibar state to use in investing in social welfare or urban service delivery – collective consumption needs simply did not appear as a priority, particularly during the individually acquisitive Amour regime. Zanzibar's current President, Amani Karume curbed these excesses some after coming to power in October 2000, but he also increased the rate of revenue flow to the mainland from the islands.

The key importance of the political strife that has ripped Zanzibar apart in the past ten years for the expansion of neoliberalism is this: it resulted in the collapse of what had become, in the late 1980s and early 1990s, a fairly significant flow of development assistance directly to Zanzibar from the West. The largely Northern European aid flows of that era heavily emphasized environmental issues. For instance, the Commission for Lands and Environment had a budget in the early 1990s in which more than three-fourths of its donor aid went to its Department of Environment. With the exception of the UN and some private agencies like the Aga Khan Trust for Culture, the donors froze aid in late 1995 to protest the manner in which CCM had prevailed. The freeze lasted from that first election until the middle of 2002 for many donors. A huge Finnida program for land management ceased operations in early 1996 and did not resume until the middle of 2003. When the Zanzibar Sustainable Program's UN money ran out right before the 2000 election, it was a year before the German Development Service stepped forward, well after the January 2001 police action in Pemba. This occurred at the very moment when donors were falling over themselves to pour money into Dar es Salaam – quite ironically, since the police forces, and the atrocities they had committed in January 2001, were a "Union Matter," not a Zanzibar government one. The longstanding ties of CUF leaders with expatriate Zanzibaris in Scandinavia – particularly Denmark, which committed to funding at this time every Tanzanian SCP office *but* Zanzibar and "where the CUF external movement started" – cannot be discounted in understanding the twisted logic of the Zanzibar-only aid freeze (Mukangara 2000, p. 46).

Since much of Zanzibar's aid money had targeted environmental planning, land management, urban services and infrastructure – all of which remained harnessed to the state throughout the whole period of neoliberal transformation, in contrast to what happened in Dar es Salaam – the city can be said to have been even more detrimentally impacted by neoliberalism than the mainland's cities. This is certainly the case for ordinary Zanzibari urbanites. Economic growth in the 1990s and early 2000s proved superficial and tangential to the main impacts of the new economy on the urban majority. Tourism produced 2,600 new jobs in the 1990s in all of Zanzibar, almost all of them "low-paid service positions in hotels" (Gossling 2002, p. 4) while virtually every upper level position belonged to expatriates. Instead, what most people experienced were the escalation in the cost of living and the depression in the value of incomes – if indeed they had incomes to report. Structural adjustment and neoliberalism's double whammy of high inflation and high urban unemployment left most urban Zanzibaris with less money and more needs

to be met with it, such that the populace defined the era as one of a serious
squeeze – rather than a broad increase in their quality of life. "There are goods
in the store now," one resident told me in 1992, "but we can't afford any of
them." In 1999, another Zanzibari explained neoliberalism this way: "Children
go to school, 200 shillings for this, 700 for that, you tell them you don't have it
and they get sent home. You feel ashamed. Even the father, he … comes back
with nothing, morning tea time comes and he tells his wife, 'Make a plan.'
Imagine that! What would *you* say if you had to tell *your* wife to 'make a plan'
for breakfast!" (Myers 2002a, p. 156). Because most city residents also identify
this era with the declining autonomy of their country within the Union, it is not
surprising that the bitterly spit name on the street for the time of *ubinafsishaji*
(privatization) in Zanzibar is *ukapa*, in recognition of the United Republic
President, Benjamin Mkapa (1995–2005).

In Zanzibar, the limited elements of privatization that have been introduced
in spheres such as electricity or solid waste management have indeed increased
the degree to which urban services are delivered in certain areas of the city.
It is the wealthiest parts of Zanzibar city that have seen the greatest
improvement in electrical supply provision, for instance. One result is the
replication of the colonial bifurcation of the city into two parts – one, the
Arab-Indian-European elite area of Stone Town, and the other, Ng'ambo,
which literally means the Other Side of town. Almost all major donor funding
for urban development in the city during the neoliberal era has poured into
Stone Town, along with nearly all of the city's new hotels, restaurants, and
tourist shops. The exception has been elite private investment outside of Stone
Town – in extensions of what were the elite zones of beachfront villas in
colonial times north and south of the city, mostly in West District. Much of
Ng'ambo and its many extensions (combined, this is where 95 per cent of the
urban population now lives) remains without sewage connection, proper
drainage, solid waste collection, efficient water and electricity supply, or
sufficient educational services. By contrast, Stone Town is visibly cleaner than
it has probably ever been, especially if we remember that this was the place
David Livingstone called "Stinkibar" in the 19th century (Myers 1994c). Its
sewers, water and electricity networks, and waste collection system are in top
form, comparatively speaking, after twenty years of donor funding and private
investment. The stink of Stinkibar in the 21st century resides much more in the
rutted alleys of Ng'ambo, particularly the farther one goes eastward from
Stone Town in the city.

Since urban service provision, land delivery, and a great many basic aspects
of everyday life in Zanzibar remain state-controlled, it would be tempting to
dismiss the impacts of neoliberalism here. Yet the facts that neoliberalism has
made the Zanzibari economy so heavily weighted toward tourism and that
tourism is concentrated in the Stone Town have meant that the state skews its
resources in this direction. To the extent that the Zanzibar Municipal Council
retains any authority to collect its own revenues – for business licenses and the
like – almost all of the revenue is obtained from Stone Town. (And even here, it
must compete with the Stone Town Conservation and Development Authority,
created with major donor funding in 1989 to oversee the preservation and

redevelopment of the "historic" area.) The Zanzibar Chamber of Commerce formed in the neoliberal era is almost entirely a Stone Town organization. With pressure groups like that and revenue collection like this, it is almost inevitable that the state's actions are generally weighted toward interventions beneficial to the Stone Town. Two of the five action plans of the ZSP have been Stone Town based, when it is an area where less than 5 per cent of the city lives and its environmental health problems are minuscule in comparison to Ng'ambo.

The solid waste collection conducted via the Zanzibar Municipal Council in the 1990s and 2000s replicated this bifurcated map. Overall, the Council's solid waste managers estimated in 2003 that in a good year they collect approximately 40 per cent of the solid waste produced in the city. In 2002, some 72 out of 200 tons produced daily were deposited at the dump, or 36 per cent. More than three-fourths of the waste of Stone Town makes it to the city landfill – and, on a good day, nearly all of it does – in contrast with less than a third of the garbage of the rest of the city. Some of the older parts of Ng'ambo adjacent to Stone Town have collection rates closer to those of Stone Town, but other wards never see the city's collectors.

Mkele was a typical, or average, ward among the 26 wards inside the Municipality, making it a good choice for a pilot project. In 2002, it produced about 7 tons of waste per day, and the city collected about 2.5 tons of that, or 35.7 per cent (ZSP 2002). Residents either burned the rest, buried it, or dumped it into the two stream channels that mark Mkele's borders. Most of the labor of dealing with waste informally, in Mkele and the rest of the city, is performed by underage children without any form of protective clothing or equipment. Air pollution, water pollution, soil toxicity, increased flooding, water-borne disease, and other environmental hazards are consequently basic features of Mkele's lifeworld, as they are for most Ng'ambo areas, and the fallout falls especially hard on children.

The renewed bifurcation of the city, in solid waste as in everything else, could potentially be altered as a result of the pilot program in Mkele. This small neighborhood about two kilometers east of the Shangani Point's beak had a little more than 6,000 people in 2002, crammed into two hectares of relatively low-lying land just north of the oldest planned industrial area of the city, Saateni (Figure 4.3). The Mkele Ward Development Committee's project, with more than US $28,000 in funding from the DED and the German embassy, aimed to collect and deposit all of the household waste in the ward, for a meager fee to be collected from each household.

Before it began doing so, however, the Ward Committee conducted a careful survey of the ward's inhabitants and ran a series of public discussion fora on their idea. They wanted to set a fee for the service that would be acceptably low enough for residents to pay, and they wanted to spend time articulating how the program would work. Ideally, the residents attending these mini-consultations would in effect set the prices for collection, help design the routes, and advise on methods of collection that would be most amenable to their lifestyles. The meetings were also to be used to create sites for the future construction of cement slabs for secondary waste collection.

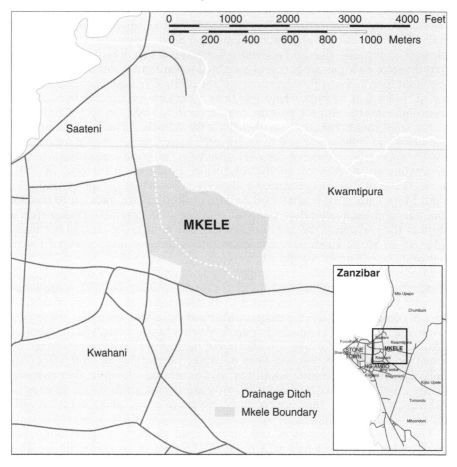

Figure 4.3   Mkele neighborhood, Zanzibar
Source: University of Kansas Cartographic Services

Out of the sensitization sessions, the Committee emerged with a plan for collecting household waste by wheelbarrows (the alleys are too narrow in most cases for any larger vehicle, to say nothing of the many other factors that make even the widest impassable by motorized transport). The ward was divided into four sections, and collections scheduled from each section every other day. Each of the eight people employed as collectors had responsibility for 124 houses, with a ninth employee assigned to collect fees, another to do the accounting, a driver, and three truck crew members. Only one of the sections agreed to a site for a slab, so that all secondary collection (meaning transport of the waste by truck to the dump) had to pass through this one slab for the first year of the pilot's operation.

I attended two of the community sensitization sessions. The driving force behind Mkele Ward's engagement with the ZSP, the man who really created the pilot program, and the dominant voice at these sessions, was the *diwani*, Kassim Juma Omar. But Kassim also made a considerable effort to listen to the wide variety of people in attendance. Even when their comments and suggestions veered far afield from the solid waste pilot program, he took note and pledged to work for a stakeholder-based solution to whatever the complainant addressed.

Clearly, there have been people involved in the Mkele ward project like Kassim who see their work as a form of mission. They have been engaged in environmental consciousness raising, political organizing, and cultural change. When I walked with the fee collector, Bimkubwa (nicknamed Bim), on her rounds – a full month before she would start collecting fees – and one young woman said she had no garbage to be collected, Bim turned into a sassy educator. "What did you have for breakfast this morning?", she asked. When the woman said she had prepared cassava, Bim started walking away, and said "huh, I guess you had some new kind of cassava without any skin … you know, we aren't collecting the fee for another month, but you should sweep better around here." The customer promptly produced a bag of garbage for the collectors. Similar incidents of teasing engagement and environmental education happened at nearly every house.

Yet there are just as many sentiments suggesting motivation for participation lies strictly in profit-making and income. Like similar marketization or privatization schemes across Africa and across the SCP map of the globe, Mkele's pilot was promoted as one that generates employment and income. Indeed, fourteen people were employed in the pilot as of August 2003. The eight collectors and three truck staff members were paid 50,000 shillings each per month, the official minimum wage (about 50 US dollars, working out to about $1.67 per day). The driver made slightly more (60,000 shillings), and the project accountant and Bim, as fee collector, a bit more still (70,000). The funding kicked in as of July 2003, but the July 2004 start date for self-sufficiency (when the fee collections were supposed to cover the salaries) had to be postponed, and at one point several employees were temporarily laid off to meet payroll. Some members of the NGO steering committee wanted to divert profits into starting up another collection zone – in one of the richest West District neighborhoods instead of Mkele, on the premise that higher fees could be charged and greater profits earned.

The Mkele ward project, with its community activism and collaboration between community groups and the local government via its version of EPM, is thought by its supporters to provide a model for the possibilities of how neoliberalism might proceed. Certainly, the manner in which the project's people interacted with Mkele's "stakeholders" in 2003 seemed to me to be something like a model for a new relationship between the populace and the governance establishment. It remains much more debatable whether the EPM process in Mkele – circumspect as it has been – will be replicated in the privatization process that Zanzibar has been undergoing. Indeed, any sort of participatory, democratized process looks very unlikely, especially because the model is Dar es Salaam. While the Mkele Ward Development Committee leaders went to Hanna Nassif to learn from Kimwoda how to operate as a community business, the Zanzibar Municipal Council was told by their Chief Minister, Shamsi Vuai Nahodha, to go to a higher power. The Chief Minister told the Council, in a stinging reprimand in the House of Representatives in August 2003, to spend two months at the feet of Charles Keenja, and to take their game plan directly from his dictatorial City Commission in Dar es Salaam in the late 1990s (Mohammed 2003, p. 19).

During that two-month period, the Chief Minister launched an emergency campaign, enlisting the National Service, Volunteer Service, Coast Guard, the Fire and Emergency Brigade, and work-camp prisoners to clean the city (*Zanzibar Leo* 2003a, p. 1). Nahodha forced the Municipality into writing a new by-law for privatizing solid waste management, with language lifted directly from the by-law for Dar es Salaam, in addition to a by-law forbidding haphazard construction (the means by which almost every new Zanzibari home is constructed) (*Zanzibar Leo* 2003b and 2003c). He gained great political capital from his decisive statements amongst a populace that "appeared to have wearied of the Zanzibar Municipal Council" (Mohammed 2003, p. 19).

With the passage of the new by-laws and the opening of the market for waste collection, Zanzibar's director of solid waste predicted that all of the flatbed trucks seen sitting idle with construction material would switch to garbage collection for their profits. This remained to be seen as of 2005. It is possible that a Keenja-style, or Nahodha-style campaign of privatized collection under duress, after a militarized cleanup, might – at least in the short run – serve to reduce the huge backlog of wastes that go uncollected in the city. What it will not do is to build anything like the participatory and democratic governance framework idealized in the EPM gospel and internalized by the gospel's converts in Mkele, or elsewhere in Zanzibar city.

Sustainable Development in Zanzibar

Again, as in the Dar es Salaam case, the environmental rationales for the attempt to implement that EPM gospel are substantive. More than 60 per cent of the Municipality's solid waste, and virtually all of the waste generated in West District suburbs, goes uncollected. This uncollected waste directly and indirectly contributes to an inter-related set of environmental health crises besetting the city. Many of the poorest neighborhoods lie in depressions unsuitable for human occupation that are subject to periodic flooding and the presence of standing water. Many of the West District squatter areas are on steeply sloped hillsides that are rapidly eroding away in the rains. The only forest reserve in Urban-West Region, Masingini Forest, the catchment zone for most of the springs that provide the city's piped water supply, increasingly has been invaded by settlement. Combine "careless disposal" of industrial waste with the above, and Zanzibar has a potent recipe for bacteriological water pollution and water-borne diseases in most of its neighborhoods (ZSP 1998, p. 19). There is no possibility for sustainable development without an integrative effort to work toward solutions to this web of problems. Via EPM, the ZSP would, theoretically, represent a program for producing those solutions.

Before the Mkele pilot, indeed, even before the Zanzibar Sustainable Program was formally recognized, the Commission for Lands and Environment attempted to utilize the EPM framework to address this complicated set of issues in a different pilot program in two West District wards. The first, Mto Upepo, was an informal settlement formed in 1993 by some 140 people on the northern edge of the city, around the key spring for the city's water supply. The second, Kijito Upele, was the neighborhood into which the government sought to resettle the Mto Upepo residents. The long rains of 1994 triggered substantial planner concern for the security of the spring from contamination by the settlers. COLE's director for urban planning – the key person behind the creation of the Zanzibar Sustainable Program, Makame Ali Muhajir – brought the acting lands director (Said Omar Fakih), the Urban-West Regional Commissioner, West District officials, ward officials (*masheha*), and the Mto Upepo residents together to negotiate. Muhajir and Fakih sought to convince the squatters of the danger their settlement posed to the whole city, and to convince the West District officials and masheha from Kijito Upele that new plots and new residents would not be detrimental to their neighborhood's development.

Little of this negotiation process was easy. All of the settlers in Mto Upepo were of Pemban origin and were strong backers of the CUF. The fact that Muhajir, as a Tumbatu islander, and Fakih, as a Pemban, came from disaffected corners of the realm as well helped somewhat to ease tensions, despite their obvious CCM loyalties as government officials. Their jovial personalities also put the parties at ease. Still, "minor matters became political, and tempers flared all around more than once" (Myers and Muhajir 1997, p. 378). It took a year and a half to accomplish the resettlement, with fencing and greenery around the spring and a new, semi-planned neighborhood on higher ground at Kijito Upele. As Fakih (1995) put it, "a key lesson learned was that absolutely nothing in this day and age can be accomplished without working closely with the people affected by development; you have to involve them from the beginning."

Yet the future utility of this minor example of fairly effective EPM-style stakeholder involvement was directly undermined by events a few months later, a few hundred meters from the first squatter area, in another part of Mto Upepo. A major transformer at the nearby electrical power station exploded in early 1996. The revolutionary government, under siege and in the donor doghouse for its dubious electoral victory, blamed the explosion and resultant power outage on squatters who had settled within what the government produced as the security perimeter of the power station. Again, all of the settlers were suspected CUF supporters of Pemban origin. This time, there were no Mr. Nice Guy negotiation visits from Muhajir or Fakih, but a swift and immediate demolition of more than 100 homes without compensation or relocation. Exit EPM, enter FFU (the Field Forces Unit, the Tanzanian

Army's rapid response force, known by the CUF diehards as "Fira, Fira, Uondoke" – sodomize, sodomize, depart).

Such brutal strong-arm tactics had become relatively rare in Zanzibar after the first president Karume's assassination in 1972, but one of the consequences of the donor freeze was a renewed freedom from oversight for the Amour regime. "Harassment of CUF members and their supporters … in terms of dismissal or suspension from employment, denial of certain services … and even detention" continued unabated for five years (Chachage 2000, p. 99). The environmental fallout from the reign of the "Commando," as Amour's supporters called him, came in many forms. It was there directly in the cutting of clove trees for firewood all across Pemba (where 85 per cent of the Zanzibar clove crop comes from) and the dynamiting of reefs in preserved areas closed to fishing, where environmental sabotage became one of the weapons of the weak (Myers 2002a). It was there more subtly in the form of the sullen and fearful hostility that nearly every environment-related field officer faced in CUF strongholds – an apt description, by any rational estimate, for half of the wards of the city, if not more. The promise that EPM suggested for Zanzibar in the 1994–95 negotiations in Mto Upepo was lost in a haze of Commando tactics.

The increasing isolation of the Amour regime internationally – and even within the Tanzanian union – encouraged some planners "to think of what we could do ourselves, for ourselves," as one friend put it. More than this, though, it opened the door to environmental disaster. Heavy rains in 1998 brought extreme flooding, and the prolonged presence of standing water coupled with the chronic lack of clean water sources in many of the West District and outer Ng'ambo neighborhoods brought on the worst cholera epidemic Zanzibar had experienced in twenty years. With its near pariah status, and without donor money, the revolutionary government was virtually paralyzed in inactivity. Until a flow of donations triggered by Zanzibari exiles on the internet, who pooled their resources on a bipartisan basis to provide medical supplies, it looked to many people like Zanzibar was headed to a health crisis that would rival the nineteenth century cholera epidemics that James Christie (1876) made notorious.

This was the exact moment at which the Sustainable Zanzibar Program began, and it is no accident that the environmental profile produced in 1998 is rife with urgent references to cholera and its environmental triggers. Despite this urgency, six years on the city is basically right where it was, waiting for the next epidemic. Cases appeared in the city even in August 2003, well beyond the rainy season of March-May (Mussa 2003a). The solid waste dimension has changed little – indeed, collection rates have slipped somewhat in these six years (Juma 2003; R. Keis 2003). There has been some slight progress – again, as in Dar es Salaam, the dump has become a site of some recycling, and particularly of compost production via investment from a Kenyan firm in 2003.

But the garbage problem clearly persisted, even in Mkele. My field survey indicated 32 per cent of the households in the pilot collection zones still had significant unswept and uncollected waste in the vicinity of the structure a month into the collection program. Almost 20 per cent had had no participation yet in the program despite nearly two years of sensitization and

Figure 4.4   **Municipal Council garbage truck cartoon: The truck has the ZMC name on it – Zanzibar Municipal Council. The character says: 'Folks, this garbage slips or slides on to the road. Wouldn't it be better if it were left where it was?'**
Source: *Zanzibar Leo*, August 8, 2003

preparation. And areas of public space in the ward were littered with waste or evidence of burned waste. Residents of neighboring communities added to the problems by dumping their waste in Mkele, particularly in the drainage ditches.

The endurance of the waste problem completely outstripped the beginning days of the Mkele pilot's operations in both media and general public attention. The government's newspaper, *Zanzibar Leo* – begun under the tutelage of Chief Minister Nahodha a few years ago – turned a contemptuous eye on the Municipal Council at the very moment that one of its own councilors was engaged, through a public-popular partnership, in the potentially prominent Mkele pilot program (Figure 4.4). The Municipality attempted to defend itself by saying it lacked the trucks and the revenue by which to carry out comprehensive solid waste removal (Mussa 2003b). Public sentiment clearly ran against the Municipality, and nowhere more so than in the CUF's short-lived weekly newspaper, *Dira*. One of *Dira*'s brilliantly polemical essayists – one cannot say that *Dira* was ever actually reporting news – mourned for "our poor city:" walk in it, "and you will see for yourself how the rats dance the *rumba* and the *chachacha* while the flies help themselves to

their drink and party on in the dumps and drainage ditches full of garbage"
(Sanya 2003, p. 9). As the Municipality began to debate its new ideas for by-
laws to privatize solid waste collection and to prohibit haphazard construction,
Dira responded by stating flatly that

> the Municipality has no policies. This is the 39th year after the revolution and the
> Municipal Council has failed to properly dispose of household waste, let alone feces
> ... Why are we wasting time in seminars discussing the causes of cholera when we
> have the reasons right in front of us? Today in many areas like Kwamtipura,
> Tomondo, Magomeni, Chumbuni, and the like [far eastern Ng'ambo and West
> District squatter areas] it is difficult to provide access to social services given how the
> houses are sewn together. When a fire happens the fire truck can't reach, let alone the
> garbage trucks.
>
> (Shani 2003, p. 2)

The accuracy of Dira's cleverly spun zingers, and the bated breath with which
the public awaited its Friday morning arrival on newsstands across the city,
suggest the numerous ways that Zanzibar still falls far short of environmentally
sustainable development.

Good Governance in Zanzibar

Zanzibar today has an overdeveloped state. But it has had one for a very long
time. Julian Asquith (1970) was asked by an interviewer to reflect on his
feelings upon arrival as a British colonial officer in Zanzibar in 1952. He said it
was "one of dismay. I felt like a cog, not in a wheel, but in a very small watch."
British colonialism ruled the place with a singular lack of ambition for the first
fifty years of its Protectorate (1890–1940), but then devoted great attention –
particularly to the city – in the 1940s and 1950s. There were a lot of cogs in this
small watch; as another colonial officer of the era, A-P Cumming-Bruce (1952)
put it, the state used "a stupendous hammer to crack a few nuts." The Ten-
Year Development Plan of 1946–55 was the largest investment in the city, by
far, in the entire colonial era. It resulted in what are still today the city's main
hospital and high school, as well as numerous other schools, a civic center, an
airport, a mental hospital, and several planned neighborhoods (Myers 2003).
    Solid waste was only systematically dealt with in the Stone Town for much
of the early years of the colonial era, while Ng'ambo people largely had to do
for themselves until the 1940s (Mandrad 1992). The colonial regime used the
waste it collected in Stone Town to fill in the Pwani Ndogo tidal creek that
separated the elites from the Others of Ng'ambo. Then, in the 1940s, it used
trash to fill an unused quarry to make a football stadium not far into the Other
Side. The British certainly concerned themselves with environmental health,
since the Medical Officer of Health actually served as the officer in charge of
building regulations and planning mechanisms for Ng'ambo for much of the
colonial era. Ng'ambo residents recounted much stricter postwar enforcement
mechanisms for what they referred to as the Rat Office (the office of the

Medical Officer of Health), particularly under Dr. Robert Taylor in the 1940s and 1950s (Jobo 1992).

But social control was the chief interest of the colonial state in urban development and environmental planning. Colonialism tied Zanzibari Africans to their side of the city, Ng'ambo, the Other Side, as a "Native Location," not legally defined as a part of the "Town" of Zanzibar at all until the 1940s (Myers 1994a; Fair 2001). In the Public Works Department offices responsible for waste collection, the British Chief Secretary argued that the "presence of a European is essential" and should be "always felt," so little were Zanzibaris trusted to do anything in their own city (Dutton 1948). Despite its being a smaller city than Dar es Salaam with a longer relationship of closer administration between state and society, if anything the circumstances of the colonial legacy are worse for Zanzibar than for the mainland. The colonial state's general neglect of the needs of Zanzibar's urban majority African community was a seldom-noted part of the kindling that exploded into the fires of a socialist revolution in 1964.

The British amplified and codified the minimal pre-existing distinctions between the two sides of the city and cast their existence back in time, when in reality the Stone Town-Ng'ambo boundary was much more fluid in pre-colonial Zanzibar (Sheriff 2002). They heavily weighted local government toward the Stone Town's Arab and Indian minorities, ignoring that side's still significant African population. The earliest local government bodies were distinct – Stone Town merited a Town Council, while the Other Side fell under Native Administration as a Native Location with its own separate council – still less than half African Muslim, and not popularly elected – until the early 1950s. When the two bodies were united, the African Muslim majority population – some 80 per cent of the city – was represented by less than one-fourth of the seats on the Municipal Council.

The skewed logic of colonialism became increasingly clear in the era Zanzibaris call the "Time of Politics," 1954–63, the decade leading up to independence and revolution. Constituency boundaries in the city, and the Protectorate as a whole – were drawn out of proportion to population size in each of the pre-independence elections (1957, 1961, and 1963), favoring the elite-dominated constituencies of Stone Town. The Afro-Shirazi Party, thoroughly dominant in Ng'ambo, won majority and plurality victories in each election by vote totals, and yet was denied the opportunity to hold a majority in the Legislative Council at any time (Lofchie 1965). The colonial government busied itself in the city with a Zanzibarization of administration that essentially amounted to turning local government over to the Zanzibar National Party of the Sultan's Subjects (ZNP, but commonly known as Hizbu). This party's nickname came from the shortening of its name in Arabic – more than suggesting that this was a Stone Town based, Arab dominated party. Hizbu slang for the ASP, Gozi (skins) is equally suggestive of the ZNP's perception of ASP's base as being among Africans of mainlander origins. The Urban District Commissioner of the Time of Politics, a member of the Omani royal clan, Busaidy, orchestrated the collection of "coffee shop gossip" for the British senior administrators, as a means of keeping tabs with the Other Side, where Gozi predominated. Even the British Police chief of the time later admitted how horribly off-base and misled most of the colonial administration was about the depth of anti-Arab, anti-Hizbu feeling in Ng'ambo.

With the smaller, Pemba based Zanzibar and Pemba People's Party (ZPPP), Hizbu was able to claim a one-vote majority in the Legislative Council after the June 1963 polls and to lead the coalition government that the British turned over the keys to in December 1963. This none-too-subtle British maneuver shut out the ASP, which had won 57 per cent of the popular vote in the final pre-independence tally. As one resident put it to me, "you could smell the revolution coming" after the December inauguration of the ZNP/ZPPP minority coalition regime. A month to the day after independence, just after midnight on January 12, 1964, a rogue band of ASP irregulars took control of a police station in Mtoni. They then fanned out with the seized weapons to take the city. By the end of the next day, the ASP's leaders mysteriously reappeared from Tanganyika and a new government began to take shape. For the next three and a half months, though, Zanzibar tottered this way and that under a motley collection of leaders who called themselves the Revolutionary Council and seemed to make things up as they went along (Petterson 2002). Somewhere between 5,000 and 13,000 people lost their lives, most supporters or suspected sympathizers with Hizbu, many of Arab, Indian, or mixed ancestry. The chaotic 106 days of independent, revolutionary Zanzibar came to a close after the secret negotiation of a union between it and Tanganyika that became public on April 26.

Colonialism's Time of Politics, then the 1964 Revolution, and then the puzzling Union, left Zanzibar permanently scarred with division: Stone Town versus Ng'ambo, Hizbu versus Gozi, Pemba versus Unguja, and Zanzibar versus Tanganyika. The revolutionary state's ambitious programs for the city ultimately did little to improve state-society relations or to undo these scars (Myers 1994b). The social movement for human rights and multi-party democracy in Zanzibar over the past two decades has had its primary concentration of activities within Zanzibar city, but its map of activity does little to dissuade the observer from the conclusion that history is repeating itself and improving on its divisions. Although the opposition Civic United Front's parliamentary representation city-wide slipped to zero in the 2000 elections from two out of thirteen seats in 1995, this result reflected the CUF's boycott of the dubious re-run of that election in those thirteen city constituencies. Its two seats in 1995 were the two wards of Stone Town, and it surely would have won that territory in 2000.

The CUF, to its credit, is not a mere replication of Hizbu, nor is it entirely a replication of the ZPPP even though it enjoys massive support in Pemba, as that earlier party had. It isn't completely some amalgamation of the ghosts of both parties rolled together, either. But CUF also does not *only* share the geography of support of these two parties of the past. It clearly enjoys a funding base from exiled supporters of the earlier parties, and its policies are not a complete departure from those of Hizbu or ZPPP. Like many opposition parties in Sub-Saharan Africa, CUF cannot be seen as a progressive or radical alternative. Its proposed policy goal is *utajirisho* [enrichment]; its party platform has planks for the complete privatization of land; and it says it would actively seek foreign investors in productive sectors. Reading between the lines, it is clear that CUF would entertain some degree of repatriation of Arab properties nationalized in 1964. Although its top leaders are as moderately secular as those of CCM, at the grassroots CUF has the stronger ties to fundamentalist political Islam in Tanzania and the world at large. The policy planks, the histories, the geographies, the ethnicities, and the classes, then, all line up in stark contrast with one another, and Zanzibari political discourse has disallowed any moderating third force for decades.

Beyond the CUF and oppositional party politics, though, Zanzibar has had an exploding number of non-governmental organizations, community-based organizations, and cooperatives that have been predominantly based in the city. Far less formally, outside of these organizations, Zanzibaris live off of an informal sector that leaves them increasingly marginalized, but we also see them straddling the divides between rural and urban, and between local and global, in deliberate ways. This is particularly so for the two largest population blocks within the peri-urban communities, migrants from Pemba island and migrants from the Tanzanian mainland. Both groups come and go and come, temporarily lodging their presence on the landscape of "Pemba Street" or neighborhoods like Kilimani, Chumbuni, or Kwahani. The seeming impermanence of so many neighbors has led many Zanzibaris to see a complete social breakdown before them, and not a new landscape of democratic governance.

Zanzibar has thus attempted to implement "good governance" in the form of SCP-style action plans in an even worse season of multi-party politics than that which Dar experienced. Its attempt to come to grips with the relations between national, city, and neighborhood levels in trying to institutionalize the SCP agenda has echoed the problematic seen in Dar, but magnified all out of proportion (Myers 1999 and 2002b). The ZSP has always been a part of a state-led effort to find technocratic solutions to a very deep-seated political and cultural crisis. From the beginning of the ZSP, the Municipal Council's participation in its operations has been consistently undermined by several factors. First, it remains "difficult to define the areas of jurisdiction of each institution in the city" (Toufiq 1998, p. 1). The clashes and political fireworks were and are most uncomfortable between the Council and the Stone Town Conservation and Development Authority (Toufiq 1998, p. 14). The rich and well-funded STCDA has its own map of planning claims that compete with those of the ZMC. Overlapping responsibilities and authorities among these administrative bodies still left "several loopholes for some administrators to wield the power and assume functions not granted by the law" (Toufiq 1998, p. 1). Toufiq's (1998, p. 13) proposition paper for the ZSP highlighted the problem of "strong personalities who take over the functions and responsibilities of 'weaker' individuals and institutions. In Zanzibar, personality sometimes matters more than offices, position or institution." The "stronger personalities" of STCDA objected to the emphasis in the environmental profile and the city consultation on the environmental problems of Ng'ambo, and STCDA's backers hindered any attempts to prioritize Ng'ambo interests (Nnkya et al. 2000). Expatriate representatives of the donor agencies mostly live in Stone Town or the elite villa zone (in which case they still spend their social lives in Stone Town restaurants and bars) – thus further aiding STCDA against the ZMC.

Second, the ZMC has little or no revenue source of its own and therefore little autonomy with which to direct policy. Even such revenue sources as would exist accrue largely to the Union government, or the central Zanzibar Government's treasury, which then supposedly reimburses the Municipality. It is regularly shortchanged; in 1996–97, for instance, it received 55 per cent of what it was supposed to receive from the central government (ZSP 1998, p. 76). Paltry salaries and appalling working conditions in the Municipality send highly qualified personnel fleeing to the central government, the mainland, or overseas. Only eight of the 1200 employees of the municipality in 2003 were university graduates, and even they had incredibly low salaries. The highest paid employees, some with Bachelors and Masters degrees, made a little more than twice the minimum wage of 50,000 Tanzania shillings per month. In such circumstances, it is unsurprising that "government employees spend more time doing their private work" (Toufiq 1998, p. 15), reinforcing the sense that the ZMC is hapless and hopeless. And the central government seems inclined to want to keep it that way, as the ZMC becomes a readily available excuse.

Third, and perhaps most critically, many of the most pressing environmental issues in the city take place outside of the boundaries of the municipal council, in West District communities with highly diverse constituencies in ethnic, class, and political party terms. The ZMC lacks any legal authority for playing a role in enacting any action plans for peri-urban communities. Most lie within a mile of the ZMC boundary; the city's two gleaming new markets, are just across the street that forms the boundary, in West District, and many businesses seek to avoid paying city license fees by setting up in their vicinity. To its credit, the ZSP did recognize that it would be unfathomable to exclude Council representatives from any working groups or mini-consultations that would deal with West District issues, and it devoted one working group exclusively to the administrative boundary headache (combined with the STCDA conflict). But the net result is that the governmental bodies alone that demand representation on a working group for peri-urban or city center issues may include, but are certainly not limited to the ZMC, the STCDA, West District officials, Urban-West Region officials, the Ministry for Regional Administration and Local Government, the Chief Minister's Office, the President's Office, and the Zanzibar Investment Promotion Agency. Any number of other revolutionary government ministries or departments – agriculture, fishing, forestry, mining, natural resources, water, construction, and so on – might lay claim to an interest in the city's edge, or its Stone Town CBD, or anywhere in Ng'ambo where the ZSP might try to work. Just fitting any popular sector representatives into a conference room for a working group meeting becomes a challenge. The ZMC often adds little to the discussion, and if any ordinary person were invited and managed to show up, only the bravest would speak up in a room where she would be thoroughly outnumbered by government officials unused to ever hearing her voice.

For these and other reasons, the governance of the ZSP received a thorough shellacking in a UNDP-funded, Union government evaluation led by Tumsifu Nnkya from the University College of Lands and Architectural Studies in Dar es Salaam and assisted by Muhajir and James Materu. The evaluators did recognize that "the project was implemented in a constrained economic and institutional setting" (Nnkya et al. 2000, p. 1). Yet they faulted the government for not coming close to meeting its commitments, particularly to the municipal government, which faced "the most severe professional staff capacity constraints" (Nnkya et al. 2000, p.1). In contrast with what is expected of the EPM approach, ZSP was faulted for making "little effort to build on and support ongoing initiatives by communities and other stakeholders" (Nnkya 2000, p. 1). All five working groups suffered from public sector domination, and at the same time the public institutions represented on them sent representatives to the working groups that were "not senior enough" to make any impacts upon returning to their government offices (Nnkya 2000, p. 1). Little, if anything, has changed in the period since this evaluation (Pessel 2003; Hamadi 2003).

The EPM approach and its stakeholder-first rhetoric has had its defenders in the Zanzibar government, but their defense often has come with something of a controlling hand attached to it. "EPM is a good thing because it helps with problem solving but there will be problems with the process if the stakeholders are disruptive and argumentative," said the official responsible for overseeing the ZSP Steering Committee for four years. "For the EPM to work you have to take great care at the beginning of the process to ensure that the people who are a part of the process are not the kind who are just there to wreck it" (Adam 2003). In other words, stakeholder involvement works best if the government chooses the stakeholders it wants to work with and ignores everyone else: the Keenja model at its best.

Zanzibar city, at the ward level, though, still shows some latent potential for building a genuinely participatory and democratic governance framework for environmental planning and management. The Mkele NGO's interactions with citizens and engagement with the ideas the Sustainable Cities Program ideally seeks to put into play are one example of this. Kassim Juma Omar proved as natural a ward politician as any I have met; he combined an earthy ability to connect with ordinary working class citizens with the charm and savvy it takes to survive the infighting of his political party. He is a CCM partisan, but one who plainly recognized that many – indeed, probably most – of his constituents were not. He had to ride a very tricky balance between the CCM stalwarts who dominate the steering committee of the NGO and the popular sentiment for CUF he heard and felt in the alleys of his ward. He knew the NGO was neither a party vehicle nor even a Council one, and so he took many opportunities to foster its independence.

Kassim's reliance on a diverse crew, and particularly on Bim as the fee collector and public face of the Mkele project in the alleyways, exemplified this. Bim is a fiercely independent spirit of Pemban origin, a divorced mother with three children and a sharp wit. The project will sink or swim as much on how people that work for it talk to their neighbors as on how much the donors fund it. The fact that the crewmembers like her were able to give as good as they got in the alleys actually seems to ultimately give greater credibility to the program.

At the same time, the ambivalence with which the crew seemed to work for the program betrayed some of the shortcomings of the pilot as a grassroots governance experiment. Bi Rahma, one of three women trash collectors, implied as much with me one day. She said she thought it would be great to have the upper eastern area of Mkele open to collection simply by putting the truck there, parking it and filling it, rather than waiting for a slab location to be approved and the slab constructed. "But it is up to *you* to tell them that," she told me pointedly. "We are just workers, we do what we are told. The people of the project have told us they want a slab there. They have their reasons, and it is not for us to ask what they are." This was all the more astounding because of the vehemence with which she distanced herself from "the people of the project," when I – naively, apparently – had assumed she was one of the very core "people" of a project supposedly built from stakeholder democracy.

In sum, Zanzibar's semi-autonomous regime has played along with the "good governance" game rhetorically. Yet it has continued the long history of marginalizing the majority of its residents in municipal governance. Many Zanzibaris, even those most engaged in civil society organizations, feel that "municipal leaders have forgotten us" (Mohammed 2001, p. 85), and they see little or no avenues for their participation in what remains a clearly top-down urban policy arena (Andreason 2001). Even in Mkele, and even among participants and employees of the Mkele pilot, this sentiment prevails.

The Politics of Cultural Difference in Zanzibar

Understanding what has gone wrong in Zanzibar economically, environmentally, and particularly politically, requires us, as Mbwiliza (2000, p. 18) puts it, to "look more closely at the internal factors within the Zanzibar social formation that have contributed towards a continuous reproduction of attitudes and mental maps of self-perception whose genesis is to be found in such a distant past." The second Time of Politics has looked, to many people, a whole lot like the first.

One of the similarities is the depth, vehemence, and geographical expression of cultural differences. To begin with the last of these, the early 21st century has, as I have argued above, re-engineered the divisions between the Stone Town and Ng'ambo. When Chief Minister Nahodha chopped the Municipality to pieces in his House speech, the mayor, Ahmed Keis, responded by calling an urgent meeting of those he considered the most important town leaders from the public and private sectors. Thanks to the mayor's courteousness, I was invited to attend the session and listen to the exchanges. It was an extraordinary and truly valuable half-day discussion amongst some 40 people. It is indicative of how the politics of cultural difference actually works, though, to consider who was there and who said what. Nearly half of all the participants were Asian or Arab, in a city that is 95 per cent African. Only six of the 30 madiwani came, all from Stone Town or the immediately adjacent, rapidly gentrifying zone across from it that falls under the STCDA. Sixteen of the 40 people there were government officials. All but two were male, and neither of the women spoke the entire time. The longest contribution came from Ibrahim Raza, a prominent Asian Stone Town businessman with close links to CCM – Ibrahim and his brother Mohammed have well deserved reputations for fearless and bluntly honest partisanship. Ibrahim said, "Stone Town is the mirror that Zanzibar holds up for the world, and it smells," largely because of the small-scale traders who do not dispose of their trash properly in the seafront Forodhani Park. It went unsaid that the traders were overwhelmingly Africans living in Ng'ambo. The only mention of the entire rest of the city came from the Police Commander, who wondered why the city didn't ask him to impose stricter enforcement of laws and by-laws there, where "people continue to be filthy." The Other has continued to live on the Other Side for the Stone Town elite, and has continued to be ignored in the halls of municipal governance.

Other cultural differences with long political half-lives are those dichotomies that relate to people's places of origin and home grounds of political persuasion. Since almost all Zanzibaris are Shafi'i Sunni Muslims, religion in and of itself plays a less obvious role in this city, by comparison with Dar es Salaam. Chiefly, the dividing lines consist of the Pemba/Unguja, northern Unguja (Tumbatu)/southern Unguja (Makunduchi), and mainlander/islander fractures. None of these is as clear-cut as it seems, and each is enacted or deployed in intricate ways. Through his clever diatribes in *Dira* through much of 2003, Mohammed Ghassany wittingly and unwittingly dissected many of these. Less cleverly, Omar Ramadhan Mapuri's (1996) pro-CCM polemic, *Zanzibar, The 1964 Revolution: Achievements and Prospects* presented the counter-punches to Ghassany's line of argumentation. I use their works as brief examples here of how Zanzibaris "reproduce the attitudes and mental maps" of division, along one sample axis of division, mainlander versus islander. My discussion leads us toward the implications of these cultural differences for constructing participatory urban governance at any time in the near future in Zanzibar.

Let us begin with Ghassany's representation of the Union between the islands and the mainland. Tanzania's Union government was consistently represented in his writings as "the government of Tanganyika." Strident Zanzibari nationalism defined his reply to Chief Minister Nahodha's stated desire to revisit the Articles of Union. "Learned Mr. Shamsi, the question is no longer about Zanzibar being swallowed, but being unswallowed!", he declared, in a point-by-point analysis of how the government of Zanzibar had lost its autonomy (Ghassany 2003c, p. 4). Ghassany slipped effortlessly in to what can easily be read as "Hizbu mode" (i.e., Arab-oriented Zanzibar Nationalist Party rhetoric of the early 1960s), even in an essay dedicated to refuting mainlander's assessments of Zanzibari politics as simply reverting back to the old time of politics. Even while saying in this essay that there was no such thing as Hizbu-ness any more, Ghassany (2003b, p. 4) ended the piece with this:

What I want to say is that Zanzibar is indeed the bud for civilization in East Africa. Zanzibaris had the luck of becoming civilized first. To therefore say that Zanzibar cannot run free and fair elections but Kenya can should burn all Zanzibaris up.

Ghassany used the noun *ustaarabu* and the verb *kustaarabika* as the terms for "civilization" and "to become civilized": literally, "Arab-ness" and "to become like an Arab." These are commonly used words in Zanzibar, but there are other, much less loaded ways of saying them in Kiswahili that are the expected terms in mainland Tanzania, and Ghassany's point was not lost on anyone who could read the language. Elsewhere, he added to this the deliberate usage of a k in front of words that commonly begin with an h in standard Kiswahili – *khabari* instead of *habari* [news], *khususan* instead of *hususan* [especially] – to inflect the writing with Arabic speech patterns. As the frequently invoked CUF slogan would have it, there are *Wazanzibari* [Zanzibaris], and there are *Wazanzibara* [people living in the islands whose real origins and allegiances are to the mainland, or *bara*]. Ghassany's writings made a powerful and clever case, straightforwardly and backhandedly, for the Wazanzibari as different, and better, people than those who rule them – including their fellow islanders.

In Mapuri's book, virtually every cultural point Ghassany made against the Union was directly inverted. The 1964 revolution and the subsequent union with Tanganyika were "the logical outcome of centuries of oppression and subjugation of the African people" (Mapuri 1996, p. 1). The oppressors over those long centuries were the ancestors of the very people behind the ZNP, Mapuri (1996, p. 1) argued, and they were "basically Arabs and certainly not indigenous Zanzibaris," the latter, of course, being the kith and kin of mainlanders. In revisiting the narrative of the revolution, Mapuri sought a means to recast the ongoing CCM-CUF dispute as a recurrence of the first time of politics, with CUF playing the role of Arab oppressors and the Union and CCM being the African liberators. Had CUF been allowed to win the 1995 election – and Mapuri would undoubtedly have made the same claim of the 2000 election – the "black (African) administration would have given way to non-black (white or Arab) rule" (Mapuri 1996, p. 59). Since CUF was "campaigning for the restoration of the Sultan," Mapuri (1996, p. 59) took this "fresh resurgence of Arab sentimentalism" for a "considerable amount of arrogance."

These competing and incompatible strains of ethnonationalism run through the underbelly of Zanzibar's cultural politics, effecting even the possibilities for EPM in the city. Cultural divisions *within* the islands have also become severely re-politicized in this second Time of Politics, although the tensions are often subtle (Myers 2000). Ghassany's (2003e, p. 4) "eternal grief" from seeing the way his culture was displayed in the 2003 Zanzibari Cultural Festival has much to do with the modest distinctions between Pemba and Unguja in cultural terms, for instance.

Yet it is certainly not only CUF voices like Ghassany that critique the residual and emergent strains of cultural practice in Zanzibar. Otherwise, it would not have been possible in early 2004 for the House of Representatives to unanimously outlaw both gay marriage and homosexual acts. The latter, astoundingly, are now to be punishable by death, in a city where the straight majority has shrugged off cross-dressing, lesbianism, and gay male sex acts for at least a century. But in the contemporary global context, Zanzibari politicians were responding to what is widely seen as western imposition of openness toward homosexuality. In the local political setting, the dominant CCM had to act to outflank the CUF, with its stronger Islamist credentials, in order to secure its culturally conservative rural base.

All the same, it is unlikely that CCM was correct to blame the CUF for the bombs that exploded in the city in March 2004. One exploded outside the house of the government mufti (Islamic cleric), whose legitimacy was questionable to Islamists because he was appointed to office by Zanzibar's Minister of Good Governance, who happened to be Christian. Another targeted a prominent local businessman. A third was defused before detonation outside the Mercury restaurant in Stone Town. Mercury, popular with western tourists, is named for the man one internet poll of Zanzibaris called the Zanzibari of the Millenium, British rock star Freddie Mercury of the band Queen. Mercury was born in Zanzibar and lived his first years there, as Farouk Bulsara, relative of the mayor at the time of independence. Mercury was openly gay and died of AIDS, making him the perfect villain to Islamists who see western culture as polluting Zanzibar. The government arrested several members of an Islamic charitable organization for the incidents, the tension from which remained through much of 2004.

All of the complex politics of cultural difference, expressed in Ghassany's or Mapuri's writings and in the acts of the House, filter down into the everyday engagement with the EPM ideal in urban governance and solid waste. In particular, the Mkele project work crew's women members faced culture in motion as soon as they walked up the hill to work. Bim's clothing choices – she often wore Pakistani-style pajama pants instead of customary Swahili city attire – were a major aspect of conversation in the alleys. Bi Rahma, a Muslim who comes from the mainland, chose to wear the same protective blue overalls as the men. Cumulatively, these seemingly minor matters made a difference in how the whole project was perceived in the neighborhood. In fact, the whole element of difference of the labor force acted upon the project in both positive and negative dimensions, particularly for the women. That Bim or Rahma dressed differently, that one is Pemban and the other a mainlander, meant that they were excused or exempted from some strictures and condemned for not following them at the same time. I asked Bi Rahma if people were getting used to seeing her collecting trash in an overall jumper:

> Ah, people are shocked. They stare in wonder. *Here.* But they wouldn't on the mainland. On the mainland, women do whatever they wish, any kind of job. There is no male or female to labor, it is just a job for a person to do. Here, they have said women need to be protected, shielded, kept at home, things like that. But … [she shrugged], to me it's a job. I work hard and I have money for the home as a result.

Bim or Bi Rahma, or the other workers on the project, seemed capable of withstanding the backbiting that accompanied their new roles in the alleys. Ultimately, they were of Mkele. And Mkele residents, in the flexible manner that endures underneath the contemporary politicized rigidities of Zanzibari culture, will for the most part accept their dress and chosen line of work as the eccentricities of otherwise okay neighbors. What seems harder is the alien culture of the project as a whole. For instance, the two German Urban Management Advisors that have overseen it were both non-Muslim, thoroughly western women (although one came from eastern Germany) who had a limited command of Kiswahili and little or no background in Zanzibari history. The whole history of German aid in the city, of which Germans involved in the ZSP are only fleetingly aware, was widely remembered by Mkele residents, as by many in the city – and it was not a warm and fond remembrance (Mwanatongoni 1991; Fereji 1992; Mloo 1964). Indeed, the donor relationship to the people of Mkele and the city as a whole has been distant, culturally scripted as a separation. Hovering, too, in many people's minds, is the idea of the whole project as another imposition of mainland policy – yet one more place where Zanzibar has been swallowed. The notion of a participatory, democratic form of planning and management brought from outside succeeding, via these agents and agencies, in such a vituperously contested cultural landscape, is rather farfetched and naive.

Conclusion

The mirror that Zanzibar holds up for the world is much different than the city Zanzibaris see in the mirror every day. Like Dar es Salaam, Zanzibar has experienced the enactment of the four rhetorical devices central to this book, in a very short space of time, at a very high level of intensity. Neoliberal policies have pried open an economy once more strongly state-controlled than that of the Tanzanian mainland and left the city's poor majority vulnerable to high inflation, unemployment, and reduced service provision all at once. Sustainable development and good governance policy rubrics fostered something close to the opposite of their ideals, and the politics of cultural difference burned more vehemently here than in either of the other two case study cities. The Zanzibar Sustainable Program really stood little chance of fostering any sort of climate for EPM-style planning in the face of such a constellation of changes.

The EPM ideal has been a terrible misfit in Zanzibar over the last twelve years. When discussed at all, there is general impatience with EPM, because it takes too long, and it requires too much feedback and "too much transparency." This is an evident effect of the authoritarian hangover suffered in Zanzibar. A true EPM might mean it would take longer to clean a city up, but it might stay cleaner for a longer period of years. That kind of sustainability requires a wide and deep political horizon. A Keenja-style campaign, such as that led in 2003 by Chief Minister Nahodha, might achieve visible results but meanwhile it destroys any participatory or democratic dimensions. Moreover, there is a strong tendency in Zanzibar for the central government to use the municipal council as a punching bag. Despite its bluntly undemocratic nature, this tactic scores easy points with citizens, because, after all, they have never actually experienced a participatory urban democratic governance framework.

One of the Sustainable Cities program's strongest advocates in Zanzibar, the former mayor quoted at the beginning of chapter one, Abdulrahman Mnoga, saw it as making a minimal difference. On the one hand, "the very fact that the project [brought] a wide range of stakeholders together for discussion is impressive. The ZSP also raised the awareness of many officials of the importance of the urban environment" (Mnoga 2003). On the other hand, Mnoga continued, "local government reforms depend entirely on the democratization and decentralization of openness and good governance. This is what we have to confront. Everything about the project has to be transparent. There are people who simply do not want to give up power. Either through their lack of commitment or their ideology or their desire for more power, people have undermined the progress of the sustainable program" (Mnoga 2003).

Zanzibar may just be an irreparably cracked mirror. Lusaka, by contrast, has the advantage of a higher degree of possibility for the popular expression of dissent. Whether this leads it to have a more effective implementation process for the Sustainable Cities Program is another matter, one that I address in chapter five.

Chapter 5

Lusaka: The Years of the Rule of Money

These were the nineties. The late nineties. They were lean years. They were the years of each person for himself and hope only under the shadow of the gods .... The years when there was a harshness in the land that had little sympathy for the weak.

(Sinyangwe 2000, p. 14 and p. 30)

Introduction

The two case studies that precede this one both originate in the United Republic of Tanzania and the model for the future of African urban planning suggested by Dar es Salaam. That model was disseminated far and wide through the United Nations system, but chiefly to other cities in eastern and southern Africa. Given the long reach of connections between post-colonial Zambia and Tanzania – symbolized by the vast and creaking Tanzania-Zambia Railway (Tazara) that in the early 1970s finally connected landlocked Zambia by rail with the port at Dar es Salaam as a tool in the fight against apartheid South Africa – it is not surprising that Lusaka became a Demonstration City for the Sustainable Cities Program that had started in Dar. The Tazara and the SCP are, of course, not all that is shared here. Each theme of this book – neoliberalism, sustainable development, good governance, and the politics of cultural difference – takes its distinct shape in Lusaka as well, coloring the beginning, middle, and end of the experimentation with an ostensibly more decentralized and participatory form of planning and of solid waste management.

Lusaka

Like Dar es Salaam, Lusaka has been among Africa's most rapidly growing cities in the postcolonial era (Figure 5.1). By some estimates, its population had reached nearly 2 million people by 2000, even though the 2000 census found only 1.1 million of those inhabitants. Most of that population expansion has happened since Zambian independence. Lusaka began as a small railway stop, and probably would have stayed that way had the Northern Rhodesia colonial government not chosen it as the site of its new capital city. The construction and transfer of the capital in the mid-1930s contributed to Lusaka's growth,

Disposable Cities

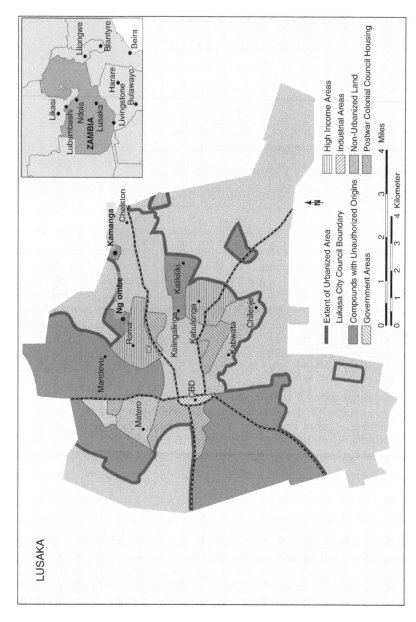

Figure 5.1 Lusaka in the early 21st century
Source: University of Kansas Cartographic Services

Table 5.1   The growth of Lusaka

YEAR	POPULATION
1928	1,879
1964	195,750
1980	538,469
1990	982,362
2000	1,103,413

Source: Central Statistical Office, Lusaka

but the city was still relatively small at independence in 1964 (Collins 1986). Colonial influx control measures and the absence of economic opportunities in comparison to the jobs available in Zambia's Copperbelt towns held the official population under 200,000 (see Table 5.1).

Lusaka has always been a fundamentally divided city, even if the divide is geographically expressed in a jigsaw puzzle of stark juxtapositions of rich and poor rather than a set of neat separations between zones a la Dar es Salaam in the colonial era. As a consequence of the past fifteen years of neoliberalism, it is becoming more of a city "broadly divided between those who do not eat and those who do not sleep" (Lubelski and Carmen 1999, p. 119). But this is a division that has existed since the city began. The colonial legacy of low density developments for the well-to-do spread across the city came with very high formal standards for service provision that were not met even in those elite neighborhoods. But they were never seen as a possibility in urban majority neighborhoods (Kalapula 1994, p. 68; National Archives of Zambia (hereafter NAZ) files LUDC 1/4/24–47; Figure 5.2).

The end of control measures and the expansion of opportunities perceived as existing for Africans in the 1960s led to a dramatic burst upwards in Lusaka's population during the first decade of independence. Its rate of growth has slowed since, but remains high. Lusaka was the only province in Zambia to register a sizable net gain in internal migrants between 1990 and 2000 (Central Statistical Office 2002). Lusaka's explosive post-colonial growth has not come in tandem with a parallel expansion of formal sector economic activity. The vast majority of Lusaka's new residents have built their homes, maintained their livelihoods, and dealt with their garbage in an ever-expanding array of extra-legal and illegal avenues.

Under Kenneth Kaunda's United National Independence Party (UNIP), which ruled Zambia from 1964–1991, the party nominally dominated much of public life in the city, including allocation of plots for building and the provision of urban services. UNIP operated as the only legal political party from 1973 to 1991, under the unified Party and Its Government (in Zambian shorthand, the PIG). The guiding philosophy of the PIG went under the title, humanism, Kaunda's term for the ideological foundations of his version of African socialism. The UNIP regime extended central government control down to local levels, to urban governments. Individual urban governments like

Figure 5.2   Air photo of the Ng'ombe-Roma boundary. Roma (on the left) was a
private, mostly white township forcibly incorporated into Lusaka
City in 1970. Ng'ombe compound (right side) formed in the early
1960s and only gained legal status as a peri-urban community
within the city in the late 1990s. This small tributary of Ng'ombe
stream that marks the boundary between the two seldom has
water flowing in it.
Source: Ministry of Lands, Republic of Zambia (1997)

that of Lusaka were, under UNIP's centralized and bureaucratized "supre-
macy" over local government, "paralyzed" almost as much as they had been
under colonial rule (Muwowo 2001, p. 57). Particularly during the 1980s,
Lusaka groaned forward under an awkward combination of increasing UNIP
repression and increasing economic hardship, through a time of "decline,
despair, and irrecoverable loss of standing" for many Zambians (Ferguson
1999, p. 12).

By the time of the ascendancy of Frederick Chiluba's Movement for
Multiparty Democracy (MMD) to power in the country's 1991 multiparty

elections, the city was already growing outside of the purview of the state far more than within it. Lusaka in the 1990s experienced the steady drone of the MMD's structural adjustment and privatization policies (Saasa 2002). On the surface, on the main roads, Lusaka's appearance changed a great deal. Gleaming new shopping malls, air-conditioned fast-food chains, and internet cafes lined up along newly widened and re-tarred surfaces. But these were clearly surfaces, and little else. The suffering of the urban majority expanded in prolonged economic crisis, in what the Zambian novelist Binwell Sinyangwe (2000, p. 30) has called the "years of the rule of money." Lusaka in the early 21st century suffers from serious air and water pollution, an insufficient water supply, failed solid waste management, rampant and myriad other sanitation problems, increasing traffic traumas, and dangerous open quarrying (ECZ 2001). But these environmental problems merely compound pervasive economic hardships that the new leaders have proven as ill-equipped to solve as their colonial and post-colonial predecessors, in a city with one of the world's highest rates of infection for HIV and AIDS.

This traumatic period of crisis, destitution, malaise, and disease during the last decade has been the context for a comprehensive attempt to overhaul the processes and outcomes of urban planning in the city through the Sustainable Cities Program. It cannot therefore be surprising to report rather minimal successes in this overhaul. When placed in a deeper context, we see that this planning agenda in the city is merely the latest in a long line of unsuccessful planning measures echoing back to the city's inception. Yet its failures go directly together with what the anthropologist James Ferguson (1999, p. 14) terms the "crisis of meaning" that overwhelmed Zambians in the 1990s. Where Ferguson (1999, p. 236) sees the "humiliating expulsion" of Zambians "out of the place in the world that they once occupied," we find the people of the capital were "in mourning for [their] country" (Nkhuwa 2002). The contemporary quadrangle of neoliberalism, sustainable development, good governance, and the politics of cultural difference that the Sustainable Lusaka Program in some ways both represents and contends within sits precisely in the midst of this crisis of meaning and time of mourning.

The Sustainable Lusaka Program

The Sustainable Lusaka Program (SLP) officially ran from 18 November 1997 to 31 December 2001. The SLP claimed to focus its work on "disadvantaged communities in order to reduce poverty and enhance overall economic development" (Mate 2001, p. 30). The SLP concentrated much of its attention on squatter areas that are locally known as compounds or peri-urban areas (regardless of whether they are actually peri-urban in location, strangely enough). After its version of the United Nations' template for a city consultation process in 1997, the SLP chose solid waste issues as the main conduit for developing its agenda for the compounds, even though it formed two other working groups – for city-center congestion as well as water supply and sanitation. It has often been difficult to separate out the problem areas identified by the solid waste and water working groups, particularly given their

interlocking magnitude (Chipimo-Mbizule and Nundwe 1997). The Environmental Council of Zambia (1997) estimated that only 8 per cent of solid waste
in the whole city made it to a landfill – most of this from the CBD and elite
neighborhoods. Basically, none of the solid waste generated in Lusaka's high-
density peri-urban areas was collected at the time that the SLP began in 1997
(Reuben 2000, p. 8). Very few of the poor in any of the compound areas had
access to treated piped water on a consistent basis. The paucity of water,
sanitation and solid waste services led one of the SLP's consultants to call
Lusaka's compounds "disease volcanoes with epidemics ready to erupt at any
time" (Sikwibele 1996, p. 41). Nearly 2,000 compound residents died in a 1996
cholera outbreak, for instance.

In 1998, the SLP began to work in the next rows of the discursive field of the
UN's Environmental Planning and Management (EPM) framework, by
building action plans out of mini-consultations in a set of pilot areas for
each of the working groups. Ireland Aid, SLP's first main donor, developed
three pilot projects for solid waste management, in Ng'ombe, Mandevu/
Marapodi, and Kamanga compounds. In these pilot areas, the donor and the
SLP officials expressed their intentions to build on community based
participation, and to coordinate projects amongst and between NGOs, public
sector, and private sector institutions. Much of this work seemed to stall
halfway out of the starting gate.

The story of the SLP in Ng'ombe and Kamanga compounds, where the most
active community-based organizations for solid waste management existed,
typifies the struggles to implement an EPM vision of urban planning in Lusaka
(Mutenga and Muyakwa 1999). Since I use examples from both neighborhoods
throughout the chapter, it is vital to situate them both, literally and
figuratively, in the beginning of the story.

Ng'ombe is a peri-urban compound that began in the early 1960s, about 9
kilometers east of the Lusaka CBD. It began on white-owned pasture lands
that previously had been leased to the Witwatersrand Native Labor
Association, which recruited Zambians for work in South African mines
(hence its earliest name on maps of unauthorized compounds – Winella, a word
formed from the acronym, WNLA). Much of the land belonged to a white
farmer, who apparently used it as grazing land for his cattle (hence the name,
Ng'ombe, meaning cattle; see Cheelo 2002). It kept these working class roots as
it spread across the hillslope on the northern side of Ng'ombe stream, across
which one comes to two of Lusaka's wealthiest communities, Kalundu and
Roma (Figure 5.2). The settlement has expanded at a bristling pace. Its 1990
census population had reached 17,288. The 2000 census lists Ng'ombe as
having 27,993 people, but the census conducted in 2002 by Ng'ombe's
Residents Development Committee (RDC) counted more than 40,000 (Tembo
2002). Most long-term residents are Nyanja/Chewa speakers of Eastern
Province origins, with a secondary group of Central Province culture groups –
a survey published in 1978 indicated that some 70 per cent of the compound
population came from Eastern or Central Province groups (Seymour et al.
1978, p. 71). Most recent arrivals to the compound, however, are northerners
that have left the Copperbelt in the wake of its economic collapse (Mkandawire

2002). Considered in combination, the findings of the 1978 survey and of my own research suggest that Ng'ombe's residents are long-term urbanites – more than three-quarters of the residents surveyed in both cases moved to Ng'ombe from another compound or another urban area in Zambia.

Like most Lusaka compounds, Ng'ombe has severe water shortages and yet significant problems with flooding and erosion (Kamanga 1989). In a Lusaka City Council study commissioned by Ireland Aid as a run-up to the SLP, the consultants saw a compound that was so poor (75 per cent of the households lived on a dollar a day or less) that it offered only a "limited scope" for upgrading. "A possible alternative was relocation," the authors of the study concluded (Simwinga, et al. 1997, p. 3). Instead of relocation, though, Ng'ombe became fully legal in 1998. Since then, it has gained a Norwegian built health clinic and a Housing in Urban Zambia (HUZA) community center – as well as a Japan International Cooperative Agency (JICA) –built primary school. Even with legality and donor attention, Ng'ombe suffers from a host of crises at once. Although it now has a mini-bus line serving its main market, most residents walk to work, typically across the polluted and snake-ridden marshes that surround Ng'ombe stream. Erosion on the steeper slopes of the newest parts of the settlement frequently takes away the foundations of the houses. In 2001, local leaders estimated that the compound had three or four armed robberies per night (Mkandawire 2002). When the SLP came to Ng'ombe, none of its solid waste was being collected for removal from the compound (*Insight* 1999c). Two community-based enterprises (CBEs) formed out of the SLP's mini-consultations in the compound, the Zaninge and Kwawama waste groups, began operations in 2000, but had nearly folded by early 2003 (Mundia 2003; Tembo 2003).

Kamanga compound is, like Ng'ombe, a compound where government assistance was technically impossible for several decades, because of the settlement's illegality. Kamanga began in 1968 as what Lusaka residents term an overspill from new legal neighborhoods developing around it – the Kaunda Square development to its west and, to a lesser extent, Chelston Township to its east (Figure 5.1). Like Ng'ombe, Kamanga was once part of a white farm. Foxdale Farms' caretaker, Museteka, had a young nephew whose popular renditions of Eastern Province traditional dances led to the new overspill area taking his name, Kamanga (Banda 2002). It lies several kilometers due east from Ng'ombe, 18 kilometers from the CBD and two long kilometers on a brutal road from the nearest mini-bus stop. While a plurality of employed Ng'ombe residents work in Roma, Kalundu, or other nearby neighborhoods and are thoroughly urban residents, large numbers of Kamanga residents work on commercial farms farther north along that brutal road, and a fair number are relatively recent migrants from the countryside. In Lusaka's compounds as a whole, some 48 per cent of the work force walks to work; obviously, since Kamanga has no mini-bus service, that percentage is much higher, and some farm workers walk between seven and ten kilometers each way. Eastern, Central, and Southern Province cultures (Chewa/Nyanja, Nsenga, Ngoni or Tonga) predominate, but with a slight twist – there are quite a few Mozambican-origin Chikunda families who settled in Kamanga with the closure of the Ukwimi refugee camp in Eastern province in the mid-1990s (Phiri 2003).

Kamanga is much smaller in area and population than Ng'ombe. A survey in 1988 found 3,571 residents, and the Residents Development Committee (RDC) estimated more than 6,000 people fourteen years later (Ng'andu et al. 1988). Kamanga may have begun as an overspill, but it is itself now said to have at least three distinct overspill areas of its own. Ireland Aid invested in upgrading Kamanga a decade before it had even been legally recognized, but it still remains even more underserviced than the larger Ng'ombe compound. It is also home to the third community-based enterprise to grow out of the SLP initiative. This CBE, registered as the Samalila Ukhondo (Let's Clean Up the Community) group, resembled those of Ng'ombe in its business plan, and it began its work in early 2000 (Mbewe 2002; Phiri 2003). Ireland Aid had counted 1,333 households in the compound, and Samalila Ukhondo targeted this entire number in its early ambitions. At its peak, the group managed to provide some solid waste collection for 600 households, but by the end of 2002, the number of participating households had dropped below 200 and showed no signs of swinging upward again.

The SLP's solid waste initiatives carried out by groups like Samalila Ukhondo, Zaninge, and Kwawama were supposedly channeled into a new program in 2002 within the Lusaka City Council. This project was funded by Danida, but headed by the SLP's former director. The Swedish International Development Agency (SIDA) took up the challenge of developing geographic information systems in peri-urban areas, out of the inspiration of the Sustainable Cities Program's ideal goals for Environmental Management Information Systems. Ireland Aid, at least as of January 2003, had approved a governance program aimed at developing a Development Coordination Unit in the LCC to do the work SLP had left undone in the sphere of community-city relations (Mate 2001, p. 2; Mate 2002; Kabuba 2002). The local branch of an international NGO, CARE, had purchased a flatbed truck allegedly to be shared between nine waste collection groups in peri-urban compounds, including the three surviving groups in Ng'ombe and Kamanga and six groups that CARE itself created in other compounds.

Yet these surface elements of continuance and sustainability to the SCP vision of a sustainable city mask serious shortcomings in reality. Donor funds may have continued to flow into the top of the city for somewhat related projects, but the bottom of the agenda stands on rather wobbly legs, especially if it is analyzed in the light of the EPM ideals. Out of six community-based enterprises that the SLP began to foster in 1998, the three in Mandevu/ Marrapodi existed on paper only and the other three were "limping along" by the end of 2002 (Kabuba 2002). Two of these, those in Ng'ombe, had had the financial support of the local city councillors (until 2001) behind them in gaining their first contracts. The one in Kamanga had identified start-up capital after three years in operation, but its customer base had, as we have seen, dwindled to less than one-sixth of its original list (Mbewe 2002; Phiri 2003).

The outcomes in the Sustainable Lusaka Program's model areas like Ng'ombe and Kamanga echo much of what we have seen in Dar es Salaam and

Zanzibar. Like the other two case study cities, Lusaka's experiences with neoliberalism are the first of several critical contexts shaping the policy outcomes. As in the other settings, it is crucial to see neoliberal policies within the historical arc that came before them, and hence it is to this larger policy history that I now turn.

Neoliberalism in Lusaka

As is the case for the Tanzanian cities, in Lusaka it is impossible to understand the impacts of neoliberalism in the city without first understanding its expansion into the Republic of Zambia. Like Tanzania, Zambia had several decades of socialist-oriented economic planning after independence that set the country up to be a contested terrain for structural adjustment's imposition. But unlike Tanzania, Zambia had experienced more than seventy years of the fairly large-scale capitalist urban development of mining and industry prior to the adoption of socialist humanism in the early 1970s.

Zambia exists on a map because of a strategic land grab by the British South Africa Company (BSAC) in the 1890s. Cecil Rhodes, the BSAC's director, dreamed of painting the map of Africa pink with British influence from Cape Town north to Cairo, passing through Zambia – in the colonial era, the territory was, after all, Northern Rhodesia. The name of the main North-South street in Lusaka's CBD reminds the visitor of Rhodes' ultimately unfulfilled dream: Cairo Road. Lead and zinc started the colony's mining run, but it was copper that became its entire basis for existence by the 1920s. The BSAC relinquished control of the territory of Northern Rhodesia to the Crown in that same decade, and from then until now, virtually the entire fortune of the country has risen and fallen with the price of copper and volume of tonnage exported.

Despite some fairly contiguous years of prosperity for the copper industry, at independence in 1964 it was evident that little of this wealth fed into a rising quality of life for the country's poor majority. This was especially so away from the Copperbelt towns that dominated the mining industry. In Lusaka, housing shortages manifested the gap between development at the top of the economy and underdevelopment underneath. The majority of colonial Lusaka's African housing areas were tied to employment with the white landowner; they were farmers' and contractors' compounds, in the local parlance. But employers never provided nearly enough housing for their employees, and as a consequence by the 1940s quite nearly a majority of Africans in Lusaka lived in what came to be labeled the "unauthorized compounds." These were either areas where the populations far exceeded the employment base of a particular contractor, or where those outside of formal employment's strict boundaries resided. Many landowners, both Asian and white, engaged in what was derisively termed "kaffir farming," by illegally renting plots to squatters on their land (Martin 1975).

In order to redirect the wealth of the copper sector toward the country's majority and toward the broader needs of national development like housing in

Lusaka, the Kaunda regime nationalized the mines in 1970 and consolidated them under a government parastatal, Zambia Consolidated Copper Mines (ZCCM). Most of the trained expatriate staff that had run the mines left with ZCCM's formation, leaving the parastatal with a technical expertise deficit that took years to overcome and that is sometimes blamed for reduced efficiency in production. At the same time, the 1970s brought a decline in the price of copper, violence and civil wars in five of the countries that border Zambia, and a steep rise in the price of oil (Mphuka 2002, p. 2). The Gross Domestic Product per capita collapsed, cut in half between 1974 and 1980 alone (Abrahamsen 2000, p. 87). In order to continue spending on the social development programs that the Kaunda regime had initiated to overcome colonialism's inequities, the government had to turn to foreign loans, and its external debts skyrocketed. The path toward repayment of this debt passed through the doors of the IMF.

In the early 1980s, the IMF set Zambia on a course for reduced government spending, reduced government involvement in import and export controls, and general liberalization of the economy. As in Tanzania, the world's major financial powers had little time for or interest in the prevailing explanations of the Kaunda regime for the economic crisis. Typically, these centered on the collapse of copper prices and the "consequences of supporting the struggle against [the] UDI [Unilateral Declaration of Independence]" by (Southern) Rhodesia's white minority, along with the struggle against apartheid and the civil wars in neighboring Angola and Mozambique (Zukas 2002, p. 173). The second phase of structural adjustment programming after 1985 in particular is widely seen to have had stunning consequences: "the impact of the adjustment package on the Zambian economy and society was nothing short of devastating" (Abrahamsen 2000, p. 88). The Kaunda regime, which had been reluctant to impose the IMF's policies in the first place, reneged on the SAP and made up its own version when massive riots erupted on the Copperbelt in 1987 (Mphuka 2002). Kaunda's defiance saved face at home politically, but led to a drying up of the loans and aid that kept the whole economy afloat. The IMF returned, this time with the World Bank in tow, in 1989, and structural adjustment policies came back with a vengeance.

The SAPs hit the urban poor and working class hardest, since the adjustments ended currency controls and price subsidies on staple foods simultaneously and the overall economic crisis had made urban unemployment pervasive. Adjustment-related work stoppages were a regular feature of the late 1980s, often led by the Mine Workers' Union, and large, adjustment-induced riots erupted in Lusaka and the Copperbelt again in 1990. The June 1990 riots, in which thirty people died in Lusaka, are widely seen as the beginning of the end of one-party rule and of the Kaunda regime (Abrahamsen 2000, p. 91).

Kaunda acceded to demands from above and below for a return to multi-partyism, and in October 1991 Zambians went to the polls. Chiluba's MMD won overwhelmingly because of popular frustration with Kaunda's UNIP regime for its economic policies more than any other issue. The MMD promised a great deal to its massive following, foregrounding plans for "improvements in housing, social welfare, health, and education" (Abrahamsen 2000, p. 100). What the Chiluba regime created, instead, was a full-throttle endorsement of free-market capitalism and open-arms embrace of every policy the IMF and World Bank suggested (Mphuka 2002). MMD "was the darling of the international community," the vanguard of the "new world order" of George H.W. Bush. The first Bush government joined the IFIs and the Conservative regime in Britain in explicitly endorsing MMD as a harbinger of changes they hoped would sweep the rest of Britain's former territories in Africa (Abrahamsen 2000, p. 105).

Chiluba set about privatizing as much of the economy as he could, claiming he would "privatize everything from a toothbrush to a car assembly plant" (Chiluba, in Ham 1992, p. 41). Council housing in the city was sold for nothing remotely close to its value and without any conscious planning for social equity (Schlyter 2002). New foreign aid and investment soon made Zambia one of the leading recipients of both on the continent. Privatization of most of ZCCM's activities became the monster within the privatization policy, since it was apparently far more challenging to sell off than toothbrushes or car plants. Yet by the late 1990s, most of ZCCM went to foreign buyers. In a bitter irony, the Anglo-American Corporation became one of the main "new" investors in ZCCM's holdings when the parastatal was privatized. Since Anglo-American originated in the breakup of the British South Africa Company, the Chiluba regime could quite easily have appeared as having offered the red-carpet treatment to the returning colonial masters of Zambia. The MMD government had even more to answer for than symbolic retreats like this, though, as job losses soared and deindustrialization took the country by storm throughout the 1990s. In the decade, Zambia produced one of the worst performances of all adjusting economies in Sub-Saharan Africa with respect to the expansion of poverty (Aarnes and Taylor 2003; Seshamani 2002). Despite the frequent claims of SAP's beneficial results for rural areas, in Zambia not even farming communities benefited. Chiluba's second Minister of Agriculture, Simon Zukas (2002, p. 192), reflected on the "disastrous results" of the transfer of wealth "from the agricultural sector to the banking sector." And tellingly, Zukas noted that the transfer "was not only within the country. Barclays and Standard Chartered showed extraordinary profits" for the worst drought and famine year (1992) independent Zambia has yet experienced. Since these banks had "only a token local shareholding, almost all of these extraordinary profits left the country" (Zukas 2002, p. 192).

It should come as no surprise, then, that the urban poor and working class who had put Chiluba in power became very rapidly disillusioned. Strikes and campaigns of civil disobedience began almost as soon as the MMD showed its true, neoliberal colors. One of MMD's early backers in Livingstone, Trywell Tembo (1996, p. 57) despaired over the "unfulfilled promises" of the new regime: "The MMD came to power largely due to the help of workers through the ZTCU [Zambian Congress of Trade Unions] ... and nothing is heard about poverty .... What we hear so much is about privatisation, a scheme which has no immediate benefits to the masses." Less than two years into the Zambian version of the new world order, the new regime had to impose a state of emergency to quell the unrest (Ihonvbere 1996, p. 229). The MMD coasted to victory in the 1996 elections – but only because UNIP (still under Kaunda) and seven other opposition parties boycotted the polls to protest the restrictions on civil liberties Chiluba had imposed (Baylies and Szeftel 1997; van Donge 1998). Despite the embarrassment that the second Chiluba administration (1996–2001) proved to be in any sense of the term to its creditors and the country it served, it continued to rest in the good graces of the IFIs and it continued to impose its neoliberal vision on the country. Economic liberalization continued despite a "disastrous socio-economic record" and a deliberate "complete stop" on government cooperation and consultation with civil society (Rakner 2003a, p. 16). The MMD gained a highly questionable, bare plurality victory in December 2001 presidential elections (Rakner and Svasand 2003a and 2003b). Levy Mwanawasa, who replaced Chiluba at the top of the MMD ticket only after massive internal opposition prevented Chiluba from seeking an unconstitutional third term, earned 28.6 per cent of the vote to Anderson Mazoka's 26.7 per cent (Bull 2003, p. 87).

The third MMD administration has brought only very slight changes in the direction of the country in economic policy terms, despite "elegant political rhetoric" that suggests a return to the Kaunda-era interventionist state (Aarnes and Taylor 2003, p. 169). Zambia quietly crept into the fold of countries accepted into the World Bank's Heavily Indebted Poor Countries (HIPC) initiative for debt relief, in December 2000 (Mphuka 2002). Mwanawasa's Minister of Finance and National Planning Mg'andu Magande has repeatedly stated his resolve to see the country through to the HIPC "completion point" of reforms that might bring as much as a 90 per cent debt cancellation (J. Banda 2004, p. 5). But the IFIs and donor countries have been only moderately satisfied with the degree and pace of privatization under Mwanawasa. The new administration failed to reach the completion point by the 2003 deadline, and the President claimed he wanted to be "selective" about liberalization (*Post* 2004, p. 3). Lack of a monitoring system for the country's Poverty Reduction Strategy Paper (PRSP) left the British Department for International Development (DFID) ambivalent about its future support for Zambia, the scale of which was said to depend very much on progress toward "public sector management reforms" and "private sector growth" (Nsama and Kaswende 2004, p. 5). The IMF complained about "extra budgetary spending" outside of the liberalization priorities – much of it targeting health, education, and social welfare (Chonya 2004, p. 1).

Neoliberal policies do not play well in Lusaka's compounds. The anti-debt NGO, Jubilee-Zambia, captures the spirit of popular resentment of the "radical privatization program under structural reforms" for its failure to "address the economic and social impacts on the poor and vulnerable people," including "massive job losses through retrenchments, ... [and] user fees in schools and hospitals" (J. Banda 2004, p. 5). Neoliberal IFI planning led to the imposition of the infamous Pay As You Earn (PAYE) system for taxes on wages. Ostensibly a redistributional income tax system, in practice PAYE meant that the "tax burden shifted from the corporate sector towards an estimated 400,000 citizens" in formal sector wage employment (Namoonde 2004, p. 7). Admittedly, since most compound residents are not formally employed, PAYE's impacts there were minimal, but that hardly meant compound dwellers had no additional belt-tightening as a result of neoliberal privatization of services.

In many ways, the title of Mulenga Kapwepwe's 2004 play, *Like Choosing between Eating and Breathing*, speaks volumes about the Lusaka urban majority's relationship to neoliberal policies. At one point in the play, which concerns the struggles of female sex workers in the shadow of the AIDS pandemic, a sex worker about to enter into a transaction with an elite businessman turns to the audience and says:

> This [Finance Minister] Magande should take us where he goes to talk with these people, [back to the] World Bank [and the] IMF International Money Funding mu Chizungu [in the lands of the Whites], am I right? Us we would tell them what they are doing to us women with their policies, [like] HIPC? ... They are creating a Highly Indebted Prostitute Community! Eeeh, street prostitutes, office prostitutes, school going prostitutes, children of prostitutes, mothers of prostitutes, prostitutes who are mothers, wife prostitutes, girlfriend prostitutes, and not forgetting political prostitutes.
>
> (Kapwepwe, in Djokotoe 2004, p. 15)

The white dominated Zambia Association of Manufacturers may want tax relief and export support in the mode of structural adjustment, but the vast majority of compound dwellers would offer a laugh of recognition at the characters' lines here (Mupuchi 2004, p. 11). The Jesuit Centre for Theological Reflection estimated the cost of the "basic needs basket" of food alone for an average Zambian family in Lusaka to be 456,300 Zambian kwacha in 2004 (about 90 US dollars). This is only slightly below the typical monthly wage of unskilled labor in the formal sector of the Lusaka economy, meaning it is virtually impossible for most families dependent on lower, informal sector wages, to meet even basic food needs, let alone shelter costs, water, clothing, school fees, or any other basic needs. In such a context, privatization of access to basic services means that most people go without them or seek alternative means of obtaining them. "We can't afford to bargain for the price of anything. We are beggars," as one Ng'ombe woman said (in Simwinga et al. 1997, p. 18). People eat less. Since water is a huge cost, bathing is minimal and certain family members are prioritized. The density of occupation in one-room units in compounds has moved beyond the imaginable numbers of people who might fit in such units (Malamba 2004).

One of Zambia's sharpest editorialists, Azwell Banda (2004, p. 10) of the *Post* newspaper, summed up the situation in the lingering crisis years of the rule of money: "The economy is still in the doldrums; our people continue to die in droves from hunger, joblessness, poverty, and HIV/AIDS. Mwanawasa's government continues to be a weak, ineffective manager of IMF, World Bank and donor neo-liberal social and economic prescriptions, to our national detriment." While Banda may be prone to fits of hyperbole, there is simply no escaping the pall of despair that hangs over Lusaka compounds in the wake of such insurmountable problems and the spider web of failed solutions to them.

The Sustainable Lusaka Program was a product of the second Chiluba administration, an administration that explicitly claimed that it wanted to "end the dependency culture in Zambian society" so that people "no longer looked to the state to provide for their livelihood" (Abrahamsen 2000, p. 133). The SLP breathed its last breaths during the new Mwanawasa regime, but, crucially, under a new City Council without a single MMD member. As a consequence of these bookends, the program combined a vigorous commitment to a privatized vision of Zambia with an alternative mindset that, for a few of its personnel, still had a whiff or two of Kaunda's philosophy of humanism to it.

The neoliberal assumptions of the SLP's work in the compounds are made plain in its renaming of community-based organizations as community-based enterprises. These CBEs were required to produce business plans and to register not with the registrar of societies but with the registrar of companies (Ng'ambi 2002). Some of the CBE leaders took to the business orientation more than others did. Some officials with ties to Samalila Ukhondo, for instance, found it generally more profitable to utilize the flatbed truck of the Ireland Aid squatter upgrading project to take people to funerals – a regular occurrence in a compound so devastated by AIDS that roughly a fifth of its children are orphans – than to engage in secondary collection of solid waste. Their business plan gave them responsibility for primary collection of waste to slabs and midden boxes scattered around the neighborhood – from which the waste was supposed to be collected by city council trucks, in their understanding. Hence even when they had a truck and petrol to run it, they rented the truck to bereaved families instead because their business made more money that way (Phiri 2003). By contrast, the leaders of Zaninge and Kwawama waste groups, neither of which had access to a truck for that purpose, also seemed to start their whole operation from different premises. "Look," said Peter Tembo of the Kwawama group, "we have built our houses here, this is our place. People simply need to understand that this is where they live, they have to be responsible for it. What we do is sensitization. We keep going, really as volunteers most of the time." Ruth Mundia of the Zaninge group seconded Tembo's decidedly non-capitalistic idea of an "enterprise" in their community: "a lot of the time people are questioning us, harassing us, you know, like we are taking the project money, we are eating donor funds to enrich ourselves, and I say, come, look at our books, look what we do." And when they did, as I did quite literally, what they saw was an enterprise that was not making a profit. Tembo and Mundia, it seems to me, were social workers engaged in community organizing, not small businesspeople.

Regardless of the outlook in the business plans and in their implementation, it is quite evident that neoliberalism has done little, if anything, for solid waste collection in the compounds. The 2002 shift to the new Danida program brought a startlingly straightforward privatization of solid waste collection (Kaindu 2002). The new project had no lines of connection to the work already undertaken by the CBEs in the SLP. The new project is another manifestation of the externality and donor-driven character of much of the supposedly participatory agenda for solid waste management in Lusaka. One of SLP's former directors, Litumelo Mate, took over the Danida office. The program was supposed to be city-wide, rather than area-based. Danida divided Lusaka into 12 zones based on rates demarcation boundaries and invited bids for private contractors for waste removal (Mate 2002). Yet these 12 zones actually excluded the compound areas in which 75 per cent of the city resides. These came under a separate plan, whereby central-government-appointed Neighborhood Health Committees were supposed to collect fees, arrange primary trash collection, and pay a separate private contractor for secondary removal.

The decision to work with the District Health Committees ostensibly came about because "they are less political than the RDCs," who often quarrel with the City Council. This program supposedly "deemphasize[d] the business component and the enterprises of the SLP" (Mate 2002) but seemed instead to turn to larger formal capitalist enterprises. Even in the compounds, the program's director noted, "local communities must understand that there are costs they need to cover" (Mate 2002). It began with five peri-urban compounds as an experiment – five areas that were not at all involved in SLP. Community participation seemed quite frankly to be a cost-cutting measure, and not in any sense a strategy for a grassroots democratization of planning. The program was "using the community for certain functions to lower the price" (Mate 2002). The program costs became even more of an issue when the newly elected government in Denmark changed priorities and slashed the program budget in 2002. As a consequence, many SLP veterans feel the Danida plan has a serious problem of "ownership ... Will it live beyond the donors, will it be linked to normal council work?", one wondered aloud (Mate 2002). Another felt that Danida's approach left "the CBEs of the SLP out in the cold and dispirited" (Kabuba 2002). That place, out in the cold and dispirited, is a place long assigned to Lusaka's poor majority in the compounds, and the neoliberal elements of the SLP seem to have succeeded very little in recreating it as a spirited heart of a participatory planning process.

Sustainable Development in Lusaka

Like most Sub-Saharan African countries during the past two decades, Zambia has developed a fairly broad legislative and institutional landscape for environmental protection. The Kaunda regime adopted a national conservation strategy in the mid-1980s, and an environmental pollution control act in 1990. The Chiluba regime expanded the commitments of the Zambian government to the environment in legal and administrative terms, most particularly by establishing the Environmental Council of Zambia in 1992 and developing the National Environmental Action Plan (NEAP) in 1994. The NEAP identified water pollution, inadequate sanitation, and air pollution as priority issues to be addressed, thus underscoring the urban character of pressing environmental problems.

In the case of Lusaka at least, physical geography or sub-surface geology can be said to play a role in the potential for environmental catastrophe stemming from the sanitation and waste collection problems identified at the Sustainable

Lusaka Program's City Consultation. The hydrogeologist Daniel Nkhuwa (1996, p. 251) notes that the poorly drained karstified marbles that "form the unconfined aquifer from which the city draws most of its potable water" make groundwater contamination quite likely given current waste disposal practices. A cynical geological observer might suggest that a city of nearly two million people has no business being located on the physical landscape that is Lusaka. The colonial regime of the British South Africa Company even thought so. It allowed its handful of white residents the right to form a Village Management Board in 1913, but there were those in power in Livingstone, the colonial capital at that time, who felt that Lusaka had "been proclaimed too soon" (Jameson 1913, in Williams 1986, p. 76) because it was such a poor site. They allowed the Board to be established only because "sanitation needed to be controlled" (NAZ: Index to files LUDC 1/1–1/25). Yet the "sleepy" and "stagnant years" (Sampson 1971, p. 44) of early Lusaka (1913–1930) came with little or no planned development or local government action to improve the site. Poor sanitation, frequent flooding, inadequate water supply, and ineffective solid waste removal plagued the town from its origins on, and Lusaka has never recovered from the slow start in meeting any of these challenges (Cheatle 1986). The Central Business District lacked a sewerage system until 1956, so that night soil carriers trundled up the Sanitary Lanes – the alleyways – behind the banks, shops, and offices to collect human waste.

Nkhuwa (1996, p. 251), who led the water and sanitation working group formed at the city consultation, contended that "with no formal plan or strategy to manage either liquid or solid waste, current waste disposal practices have increasingly been in the ground without regard to the underlying geology." Tests of boreholes and hand-dug wells throughout the city consistently show nitrate, nitrite, and fecal coliform levels well above WHO standards, and cholera and dysentary are endemic in the city. The neoliberal policy world of privatization, user fees, and PAYE clearly adds to the potential for problems latent in the geology. A 2002 research project led by scientists from the University of Zambia School of Mines tested the toxicity of water from hand-dug wells in Chawama compound and compared this with pollution levels for the few standpipe taps available to Chawama residents. The team correlated its findings with those from the District Health Committee for water-borne diseases. All of the wet hand-dug wells in the community were severely contaminated. All of the hand-dug wells that had gone dry were located in the proximity of functioning standpipes. Water-borne disease rates were found to be astronomically higher in the households dependent on hand-dug wells for their water supply. Most hand-dug wells were located on the premises of very small plots. For households without solid waste collection services – and no such services existed in Chawama – the most frequent disposal method was to burn or bury the garbage on the premises. Almost all Chawama residents have hand-dug pit latrines on the plot. Between the latrines and the solid waste, it is virtually impossible to avoid the contamination of a nearby hand-dug well given the relative height of the water table in this low-lying area.

Consequently, the research team presented its findings to the recently privatized Lusaka Water and Sewerage Company, the City Council, and central government. They did so in order to make the case that simply through the provision of standpipes to compound residents the overall health of the community would improve dramatically. They argued that this provision ought to be made without regard to the meager capacity of residents to pay for the service, since it would more than pay for itself in improved community health. The audience response focused almost exclusively on the exciting GIS mapping technology the researchers displayed with their new computer projector. A privatized, profit-seeking water company had no need to operate at a loss in the interests of community health.

Sustainable development rhetoric – apparently underlain by bedrock as porous as the limestone under many Lusaka compounds – runs through all of the 1990s-era policy statements and institutional mission statements (Chidumayo 2002, p. 33). The Sustainable Lusaka Program is part and parcel of this broadening of environmental planning activities and rhetorical devices in the 1990s. It can, like the similar programs in Dar es Salaam and Zanzibar, point to a few improvements of the environment to cite in its analyses of outcomes. On balance, these improvements become hard to find in the maze of problems that continue. Most importantly, the EPM framework as practiced in Lusaka's pilot compounds cannot be seen as offering much hope for a future in which the environmental calamities will be more limited in scope or impact.

There is no doubt that the city consultation identified a critical environmental crisis by targeting solid waste among its key issues, given the "pile up of garbage and contamination in the streets, roadsides, open spaces and the CBD" (Agyemang et al. 1997, p. 7). As Nchito and Myers (2004) put it, "reports of this nature give the impression that the whole city was like this, and yet there were and are neighborhoods with nicely clipped lawns and not a piece of trash in sight. The more affluent can either afford to pay private firms to collect their waste or have enough space within their large yards to dig disposal pits. The more affluent areas receive adequate service provision, and it is sometimes argued that these areas could subsidize less affluent areas, since, if left to market forces alone, other neighborhoods would not receive any services at all."

Whatever solutions an EPM-based decentralized and community-oriented collection scheme might offer to Lusaka's compounds, the city as a whole, much like Dar es Salaam, still faces the larger question of deposition. The Environmental Council of Zambia closed the unsanitary landfill at Libala well after it had become a severe health hazard, having granted a three-year extension to the city largely out of what seems to have been pity for the city's inability to gain rights to any other dumpsite (Mwangu 2003). When the three-year extension lapsed, the ECZ extended the city's use of a similarly problematic and technically already-full dump at Chunga. The city council had by that point identified three new possible sites through its working group on solid waste management. All three sites lay on traditionally held land outside the city boundary, land controlled by three different local chieftainesses under customary tenure. Chieftainess Mungule in Kabwe Rural and Chieftainess Chiawa in Kafue – north and south of the city, respectively – argued that they had already given up enough land to the city and to commercial farmers (Mumba 2003). Chieftainess Nkomeshya in Chongwe, just east of the city boundary, went on record as having told the council, "if you want to lose your life you will try to put this waste here" (Mate 2002). An exasperated ECZ official told me that ultimately, the failure is a matter of "political will. If the government said, 'we will build a sealed, controllable dump site and dumping will occur in a systematic manner,' then it would happen tomorrow. This isn't where the priorities lie" (Mwangu 2003).

Do those priorities instead reside in the compounds? Over the period of my research visits, I came to doubt this. The SLP had hoped that its CBEs might come to "be used as change agents not only for solid waste but any other communication on developmental issues, as they have the privilege of getting in touch with households" (Mate 2001, p. 34; Mate 2002; Kabuba 2002; Mutale 2002). But this is not a view shared by the CBEs themselves, nor by residents I spoke with in these compounds.

In Ng'ombe and Kamanga, the 1996 cholera epidemic (which killed 61 and 18 people in each, respectively) catapulted a number of local activists to become involved in the SLP as it began. Community organizing and activism has brought development benefits to the two communities over the last decade or so. Ng'ombe has a school, a clinic, and an improved connector road as a result, while Kamanga has a community center and a clinic. Both compounds have gained new standpipes. Yet the development deficits, and their environmental consequences, are quite staggering. Ng'ombe has 12 working standpipes for its estimated 40,000 residents, and even these are frequently not running. Ng'ombe people have to line up water jugs a day in advance for access to these pipes, where they find themselves the victims of severe price-gauging by the local pump managers for the private water company. The RDC supposedly employed the managers to collect fees that it would use to pay the Lusaka Water and Sewerage Company. Life circumstances being what they are, the pump managers saw their opportunity for profit in topping up the fixed price. As a result, most residents get their water from hand-dug wells or from the brackish and putrid waters of Ng'ombe stream.

"People are hungry," Peter Tembo practically screamed one day, "they are unemployed. They are illiterate." Peter rapidly ran out of dump sites for his primary collection with the Kwawama group, with the failure of the City Council to make secondary collection runs to get the waste to the dump. The slab sites and midden boxes built with the SLP's assistance were full, uncovered, and rotting most of the time. He rented a hole in the ground to store his waste in instead, covering it regularly to at least give some measure of environmental consciousness. But that filled too, and Peter said he would "have to die of blood pressure to find a new site," let alone pay for it with a dwindling number of customers. Kamanga had, if anything, even more visible garbage strewn everywhere one looked in 2003. Customers told Mzamose Mbewe of Samalila Ukhondo that they would refuse to pay for collection when they knew it was not leading to removal of the waste: "We see where you are putting it, and we can do that ourselves for free," they told her.

In the rhetorical universe inhabited by the Sustainable Cities program, sustainable development is idealized as a means of balancing environmental protection with livelihood enhancement for the poor. Operating under an explicitly business-oriented rubric in the compounds, the community-based enterprises that the Sustainable Cities Program's branch in Lusaka (the SLP) initiated worked, largely in earnest, toward that balance. It proved an unworkable attempt. Part of the blame for this most certainly rested with the framework of governance coterminous with the SLP, to which I now turn.

Good Governance in Lusaka

Frederick Chiluba spent the early 1990s basking in the glow of the western world's love of his capitalistic, democratic rhetoric. Chiluba, after all, had remarked at the fall of the Berlin Wall, "if the owners of socialism have withdrawn from the one-party system, who are we to continue with it?" (cited in Abrahamsen 2000, p. 5). In the first elections of Zambia's second multi-party era, in October 1991, Chiluba swept into power with nearly three-fourths of the voters behind him, and western observers hailed his victory as "a triumph not only for Zambia, but for the African continent" (Novichi 1992, p. 17). When he gained a second term in 1996 with a similar percentage of the vote, no one but Chiluba and his henchmen would have claimed this as much of a victory for democracy (Rakner 2003a). When his party earned a third term despite more than 70 per cent of the people voting for a party other than MMD, it wasn't really a victory even for MMD, although at least the process had some contestation to it (Rakner 2003b). Levy Mwanawasa's regime, to be sure, did bring corruption charges against a number of Chiluba's henchmen, all the while circling around the man himself. Yet, turning again to the blunt critic, Azwell Banda (2004, p. 10), "the so-called fight against corruption continues to register zero real success, because in the first place, Mwanawasa and his MMD cannot lead this fight – they are beneficiaries of the terrible political, social and economic corruption Chiluba institutionalized in Zambia."

Despite the apparent hijacking of the democratic idea embedded in multipartyism, it is still possible to make a case that other aspects of neoliberal good governance have moved forward. The Zambian media, religious institutions, and various civil society groups have played critical roles in challenging the state throughout the years of the rule of money. The strongest of these institutions are Lusaka-based, and they have helped to shape a governance landscape in the city that is healthily contested (Mukuka 2001). Mubanga Kashoki and Stephen Mwale (2003) have shown the wide degree of political awareness in Lusaka compounds leading up to the 2001 elections as a result of access to independent newspapers (crucially, the quite oppositional *Post*) and radio stations. Non-governmental organizations – and in particular two umbrella groupings of NGOs, the Non-Governmental Organization Coordinating Committee (NGOCC) of 65 women-oriented NGOs, and the larger Oasis Forum, comprised of the whole NGOCC in addition to the National Council of Churches, the Law Society of Zambia, and various other units – were the key agents preventing Chiluba's third term.

The Lusaka City Council elections and the results of local parliamentary races in the city in 2001 demonstrate the strengths of alternative, anti-MMD political organizing. MMD failed to win a single council seat, nor did it send a single representative to parliament from Lusaka. All three of Lusaka's members of parliament after the 2001 elections were opposition women, one a former mayor (Patricia Nawa) who had jumped ship from MMD.

Is there hope for a decentralized planning and governance framework built around the EPM ideal in such a context? Quite possibly. But the immediate context cannot be understood without a longer sweep of the history of local governance in Lusaka, because that is the lens through which many Lusaka residents view any reforms. Like Dar es Salaam and Zanzibar, colonial Lusaka was a city where governance mechanisms were designed explicitly to exclude the majority of citizens. In Lusaka, the exclusivity's added dimension was a greater voice and presence for European settlers.

Lusaka had a Village Management Board by 1913, was made a Township in the 1920s, and the capital of Northern Rhodesia was relocated to it in 1935. Yet local government did not gain the status of a Municipality until 1953, and there was no Municipal Council at all until 1954 (Sampson 1971, p. 69). The colonial regime and the white settler community were together in denial that Lusaka even had an African population, for many years (Wood et al. 1986). As Tait (1997, p. 152) has written, "for the most part urban authorities and European residents simply ignored developments in the African part of town." The white settler, Richard Sampson (1971, p. 73), who served as Lusaka's Mayor in the early 1960s, wrote his biography of the city as though Africans were an afterthought in the town from the beginning. Sampson (2002, p. 83) himself wrote, more recently, and upon reflection many years later about the Central African Federation of the two Rhodesias and Nyasaland, "what country could succeed when 85 per cent of its inhabitants were given little voice and that voice was opposed to its foundation?" A very similar sentiment could clearly pertain to Lusaka's own governmental structure, as articulated in the colonial city's first Municipal Council and City Council.

From 1954–63, the Municipal Council had 13 councilors, with one being appointed by the governor and a dozen elected. By the government's own wildly under-representative figures, Lusaka's population was 81 per cent African, 17 per cent white and 2 per cent Asian and Coloured as of 1959 (NAZ: LUDC 1/4/24). Technically, the by-laws for elections in the 1950s contained no race requirement (NAZ: LUDC 1/8/4). But as of 1960, the only two non-white councilors were Asians "elected in competition with Europeans." The new body was "far from being a democratic structure. It could ... not be relied upon to defend African interests" (Chileshe 1998, p. 98).

When a City Council was created for Lusaka in 1963, with now eighteen councilors, the city was divided into six wards, with three councilors for each. Ward boundaries were drawn according to ratings, and the lines carefully manipulated the spatiality of race to prevent an African majority. Africans comprised 75 per cent of the voters, but unless racial bloc voting somehow disappeared entirely, they could never possibly comprise more than half of the Council. The poorest ward contained almost half of the voting population, while the three wards with non-black majorities, combined, had less than 15 per cent of the total number of voters and yet were guaranteed half of the council (NAZ: LGH 1/13/26). The Town Council and the white-controlled Lusaka Chamber of Commerce argued strenuously for the notion of wards to be divided by valuation, because "there is a certain fear that in certain circumstances, the European and Asian voters who comprise the great majority of ratepayers, might be swamped by the African vote, with resultant predominantly African councils. This is not acceptable" (LUDC 1/8/4). The liberal western democratic notion that there should be no taxation without representation was literally turned on its head. In late colonial Lusaka, there was no representation without taxation.

The UNIP government of Kenneth Kaunda produced little advancement on the ideal of local, decentralized democracy as a governance guide for the city. For a time, after all, Kaunda abolished the city council and the position of mayor, replacing this with the office of Governor for the city, directly appointed by the President. Some have argued that the whole "purpose of the Zambian one-party state was to ensure that Kaunda retained political power with no threat or risk of losing it to anyone" (Mulenga 2003, p. 6). Certainly, in the years before the one-party system was fully enshrined in Zambia – in other words, the years from independence in 1964 to the so-called Second Republic in 1973 – other parties contested UNIP's authority, perhaps most vociferously in Lusaka (Chileshe 1998, p. 107). Lusaka had a number of noted African National Congress (ANC) and later United People's Party (UPP) strongholds among its compounds, and they didn't really just disappear after 1973. UNIP had slogans designed to encourage membership – precisely because it had trouble drawing in members despite being the only legal party. Why else would the only legal party in the country have to promote itself by claiming, "It Pays to Belong to UNIP," or resort to what one of its earlier socialist partisans eventually came to call "rent-a-crowd charades" for Kaunda's travels (Zukas 2002, p. 155)? One opponent of UNIP's growing repression wondered if the "time was not far off before it would be a requirement to produce a UNIP Card before dying and being buried in any public ceremony" (Chileshe 1998, p. 106).

Multi-partyism and the rise of the MMD adjusted the dynamics far less than first glances may suggest. Contemporary Lusaka, for all of the claims – real and imagined – of democracy having grown in it, still bears the scars of this legacy of exclusions and differentiations. In the 2001 national elections, more than 80 per cent of Lusaka's population voted against the person declared the winner of the presidential poll. Colleagues at the University of Zambia smirked at the result that 19 people out of some 4000 had voted for the ruling party at the university polling station. The new City Council had not one ruling party councilor. The new minority government at the national level seemed bent on debilitating the opposition council at every turn. It did stop short of following the lead of President Robert Mugabe in neighboring Zimbabwe, who placed Harare's opposition mayor in jail (*Zambian Post*, January 14, 2003). At the same time, it continued a longstanding trend of the post-colonial era of reducing the revenue sources available to the Council and reducing Council's fiscal autonomy in general.

The Mwanawasa MMD government began with no mandate and no parliamentary majority. Mwanawasa was able to construct a cabinet with a few opposition heads, and a number of opposition seats switched sides in parliament such that the ruling party had a bare majority by early 2002. They still faced opposition parliamentarians, opposition parties and presidential candidates, and oppositional non-governmental organizations, and Mwanawasa frequently played hardball with each. Most strikingly, he sought to de-register the Oasis Forum in a fairly ridiculous charade in mid-2004. The Registrar of Societies, an MMD appointee, announced the Forum's deregistration and the immediate registration of another organization of that exact name, to attempt to bureaucratize the Oasis Forum out of existence. Unfortunately for Mwanawasa, the Oasis Forum happened to consist of a virtual who's who of Zambian NGOs: the NGOCC, the Law Association of Zambia, the Zambia Episcopal Conference, the Council of Churches in Zambia, and the Evangelical Fellowship of Zambia (P. Phiri 2004, p. 1). The panoply of powerful churches under the Oasis Forum umbrella is most pertinent. Given that Chiluba had had it enshrined in Zambia's constitution that this was a Christian nation, it would not stand the party in good stead for long to oppose nearly every organized church in the country. The Head Bishop of the United Church of Zambia, speaking on behalf of the Oasis Forum, said, "government should know that it is making life difficult with the public because when you shut down civil society you are simply blocking people from talking in their own country" (Siyemeto, in Njovu 2004, p. 1). Even naming a pentacostal televangelist as his Vice-President for a while (Rev. Nevers Mumba, who took the job during the first year after the 2001 election and got fired in late 2004) could not save Mwanawasa from the Christian organizations' interests in democratic fair play, and the Oasis Forum deregistration fell flat.

Despite this apparent sense in which non-governmental organizations have broadened the governance possibilities in Zambia, at the level of the local government, things do not look dramatically different than they did in Kaunda's time, or colonial times. The contemporary Lusaka City Council's dependence on the central government Ministry for Local Government and Housing and international donors for virtually all of its budget has meant that its policies and programs remain tethered to outsider and minority agendas. This has bred a substantial degree of cynicism among compound citizens about the exclusionary governance around them that masquerades as a decentralizing of democracy. Ann Schlyter (1999, p. 65) cites an informant in one compound as responding to her: "Democracy? That is expensive food in the shops and nothing in the pots." This sort of "disappointment, disinterest in politics and cynicism" (Schlyter 1999, p. 65) encapsulates the dominant attitudes one encounters in compounds, and the City Council usually receives the harshest treatment. It is fairly common to hear it deliberately slurred, as the "Shitty Council."

Oppositional politicians may have carved out strongholds in the city, but apples do not fall far from trees. By 2004, arguably the strongest opposition candidate for President within the city of Lusaka would have been said to be

Michael Sata, leader of the Patriotic Front. But Michael Sata was also the Governor of Lusaka for the PIG, responsible for the major urban improvement projects of the Second Republic. There is more than a little of the nostalgia that often accompanies the aftermath of authoritarian rule embedded in Sata's popularity. One cannot help but reach that conclusion when the other opposition party candidates to retain substantial popularity in the compounds are Criston Tembo of the Forum for Democracy and Development (FDD), the longtime chief of the army under the PIG, and Anderson Mazoka, a former CEO of Anglo-American in Zambia. On the other side, Patricia Nawa, Edith Nawakwi, and Sylvia Masebo, the three women parliamentarians for the Lusaka area from opposition parties after the 2001 elections, had all been co-opted in one way or another back into the MMD fold by the middle of 2003. In Zambia, "political practices associated with one-party rule" have actually "prevail[ed]" within the era of multi-party democracy and free market neoliberalism, even at the local scale (Rakner 2003a, p. 16).

The MMD did institute a number of changes in local government, particularly by its creation of officially non-partisan and yet elected Residents Development Committees (RDCs) for the compounds. RDCs were supposed to replace the UNIP party apparatus in advocating or negotiating for resources with the city council, donors, and the central government. They were rather quickly "surrounded by confusion" in most compounds (Schlyter 1999, p. 61). As one resident of George compound put it to Ann Schlyter (1999, p. 61), "How do they think they will be able to have a non-political election in a highly politicized compound?"

The RDCs were in fact initiated and introduced by the NGO, CARE International, and then copied by the government across the city. CARE's Zambia office ran a program in several pilot compounds (including Kamanga) in the early 1990s with funding from the World Food Program and the Canadian International Development Agency, called PUSH (Peri-Urban Self-Help). A workforce of mainly female compound residents worked on neighborhood improvements – rehabilitating roads, digging drainage ditches, and collecting the garbage – and were paid in food rations. Britain's Department for International Development took over the funding for PUSH in 1994 and attempted to redirect it toward a more participatory project (termed PUSH II) – in other words one where the neighborhood improvements were those residents themselves identified along with the means of addressing them. The DFID program was scaled up to 14 compounds in 1998 and renamed PROSPECT: the Program of Support for Poverty Elimination and Community Transformation. In the middle of the PUSH II program, DFID's Lusaka officers became frustrated with "the weakness of existing organiza-tions" in the pilot neighborhoods and formed "representative area-based organizations" with democratically elected Zone Development Committees (Sanderson and Headley 2002, pp. 251–52). Calling this a form of "guided participation," CARE won the interest of the then City Council under its Mayor, Patricia Nawa, and the RDC organizational format began.

The new City Council that took office after the December 2001 elections looked with disfavor on the RDCs, however. Since they had begun under the previous MMD City Council – even though Nawa herself jumped to Tembo's FDD in her successful run for parliament in 2001 – many of the people elected to serve in RDCs were seen as MMD stooges. RDCs kept operating within neighborhoods, but with little connectivity to the new Council, which had little interest in working with them. From July 2001 until December 2002, the City Council suspended its relations with RDCs, and even the central government agreed to a review of their constitutional rights and functions. In late 2002, the new City Council announced a reorganization plan that, in effect, side-stepped the RDCs to create Area-Based Organizations (ABOs, taking over the old PUSH II terminology) that would be elected to represent larger districts. This moved power away from smaller-scale neighborhood units, and towards geographical units roughly equal with the 30 council constituencies. The central government response was to strip all matters related to health – including solid waste management oversight – away from the Council to empower the Lusaka Urban District Health Management Board (LUDHMB), a central government institution. The LUDHMB oversees all 24 government-run health clinics, each of which has a set of neighborhood health committees that then effectively supplanted RDCs or ABOs in the implementation of a wide range of urban policies from the central government (Mumba 2003). Lusaka's new mayor in 2004 started his term with a very similar mindset to that of his predecessor, dictating down to the people precisely why their attitude toward waste management would have to change with privatization (*Times of Zambia* 2004).

The SLP liked to think of itself as an institution-based effort to create "good governance" in the "classic liberal western democratic model," considering the "totality of how government is done and what it achieves" (Lerise 2000, p. 90). Given the "interchange of ideas" that developed as a result of five years of consultations and workshops, the SLP's personnel pointed to successes in the "establishment of collaborative bridges ... [that] assisted in restoring the social trust and credibility of the Lusaka City Council" (Mate 2001, p. 7). SLP leaders claimed to have built "strong linkages amongst LCC, NGOs and the private sector ... to engage in dialogue and joint delivery of services" (Mate 2001, p. 7). They held that a "positive change of attitude" could be detected amongst "stakeholders that were directly and indirectly dealing with the program and LCC" (Mate 2001, p. 9). Through various and diverse training sessions, more than 500 people did directly benefit from SLP's activities. There were also real and direct impacts in the communities where SLP operated. The water supply system in the three demonstration areas and in two replication settlements improved a bit. Three peri-urban settlements gained standpipes and boreholes. Kamanga got funds for pit latrine construction. Community profiles for Mandevu/Marapodi, Chinika, and Chibolya compounds did create some sense of community involvement in all three areas (Mate 2001, p. 10).

Chief among the constraints that the SLP's own people identify, though, is minimal institutional support from the LCC, which was deemed to be "in need of financial surgery" (*Insight* 1999a, p. 1). In the City Council bureaucracy, the SLP terminal report concludes, the EPM process was never clearly understood, "due to the entrenched approaches that the local authorities are used to" (Mate

2001, p. 11). The report is blunt about how "ineffective" the LCC is: "during the whole program period LCC's responses to community demand were slow." The organization couldn't cope with "cross-sectoral needs of the communities," had limited funds, and "minimal motivation" to work toward SLP goals (Mate 2001, p. 11). In 2003, a billboard on Independence Avenue heading into the heart of the government district had the Mayor beaming above the caption, "It is better when your rates are paid: LCC." But very few Lusaka residents have ever paid rates consistently, and very few have any apparent respect for or faith left in the Council after years of inadequacy: "There might as well be no Council," said one resident. "It is as though it is not there. They collect some rates here and there but in terms of services there is nothing" (Mumbi 2002).

The non-implementation of policies outside main priorities – and SLP clearly seems to have been outside of those – has its most severe consequences in the peri-urban compounds. In the eyes of SLP personnel, the three solid waste demonstration areas lacked capacity, and "community expectations were too high" – they "expected to have financial resources allocated to them directly" (Mate 2001, p. 12). In the communities themselves, people would reverse these conclusions – the LCC lacked capacity, the donors' expectations were too high, and all too often the ones expecting financial rewards were city council, ruling party, and central government officials.

From the CBE's perspective, the failure to remove waste from the compound "is the failure of the city council. The city council … its … [pause] … its just hell" (Tembo 2002). For Kwawama and Zaninge groups in Ng'ombe, this meant that "if we have any kind of development we just hope the council will stay out. We can have a truck, the petrol, everything, and they will come and remove the petrol, take it to their side for some other purpose" (Mundia 2002). These two CBEs say they have failed because of the Council's failures. Yet the leaders of Samalila Ukhondo in Kamanga wanted a stronger government in terms of services (Samalila Ukhondo 2001). They desired a city government that would intervene on their behalf. They see themselves as having been created by them. Part of the reason for this sentiment in Samalila Ukhondo clearly comes from the substantial role that ward politicians have played in its development. At one point, the ward councilor even invited them to come across the road and begin collecting from Chelston, where he lived. These groups clearly have two very different visions for re-engineering the relationship between the council, the central government, and the compounds. But the bottom line is that *neither vision from below matters much at all to policy outcomes*.

It seems unlikely that a program can be truly participatory when it is donor-driven and externally derived, even if it relies on solid community leadership, as the groups in Ng'ombe do. Susan Phillips (2002, p. 148) rightly asks of "externally imposed forms of social organization" like the RDCs, ABOs, and even the CBEs of the SLP: "Do they enhance supportive relationships and networks among poor people, or are they primarily an administrative convenience to enable external agencies to interact with poor people?" Unfortunately, most of what has been done in the name of social organization by externally derived initiatives in Lusaka tends toward the latter. As even DFID's designers of the CARE PUSH II program admitted, in many cases "politicians continue to make moves to control community participation for their own benefit" (Sanderson and Headley 2002, p. 256).

Kamanga exhibits striking examples of this in numerous ways. One leader of Samalilo Ukhondo was, after all, White Kapitao Phiri, who was, as his middle name suggests, the son of the kapitao [captain] of the settlement when it formed as a compound of farm workers for Foxdale Farms in 1968. He quietly worked around the manipulations of a second local politician, who had largely made the Kamanga pilot project his own personal company. A third local elite remained very much in the picture, even while a resident of the middle class planned neighborhood across the road, Chelston. Perhaps we can still make some claim for Samalilo Ukhondo as an endogenous entity for Kamanga, but it is clearly one driven by the male political elite of the compound, even while most of the people who work for it are poor women with rural roots.

Although Kamanga compound residents do see some benefits from the CARE program (Mwizabi 2004, p. 5) – which they continue to call by the simple term PUSH despite the semantics of DFID and CARE – the food-for-work dynamic became especially problematic to the operation of Samalila Ukhondo. The CBE actually took a number of its first employees from PUSH-Kamanga, since they had waste collection experience. Some Samalila Ukhondo employees wondered where their food rations were at the end of the month. Employees who remained with PUSH, by contrast, wondered why they didn't get a salary but ended up with only little bags of beans and *kapenta* (dried small fish).

Samalila Ukhondo's daily operations became even more politicized with the run-up to the December 2001 elections, and party politics never ceased to interfere afterwards. The bitter MMD politicians defeated in 2001 orchestrated a campaign to stop people from paying for waste collection, saying that it was a program exclusively by and for the followers of Anderson Mazoka's United Party for National Development (UPND) and especially Tembo's FDD, whose parliamentary candidate, Edith Nawakwi, had defeated their incumbent. Scarcely a year later, however, tables were turned. Overspill area three in the most flood-prone zone of Kamanga had several dozen homes demolished by order of the UPND/FDD controlled City Council. It just so happened that all of the homes demolished were on plots that had been illegally sold to MMD cadres by the MMD ward leaders.

The years of the rule of money were supposed to be the years of the rule of good governance in Lusaka. In the sphere of solid waste management, the working relationships that were supposed to be remade anew from the ashes of so many exclusionary tactics that had previously masqueraded as local government in the city instead seemed themselves to be simply added to the ash-heap. The competitive party politics of an admittedly slightly more transparent political system, combined with a neoliberal economic policy context, produced more acrimony and bitterness, but not more equitable urban service delivery or participatory urban solid waste management.

The Politics of Cultural Difference in Lusaka

The 1990s marked a time of great symbolic changes in Zambia as well, cultural shifts that have reverberated into the present and through each crevasse of the SLP's work. With the 1991 election, Kaunda's trademark safari suits had disappeared from clothing shops along with his philosophy of humanism. Chiluba's clothing choices consciously and deliberately marked the new era of the years of money: double-breasted suits with matching power ties and handkerchiefs. The new style rapidly cascaded through the menswear lines of Lusaka; it shared its name with the name the Zambian media gave to Chiluba's economic policies: "new culture" (Hansen 2000, p. 92). It did not take long, though, for the new-culture style to fade, and for the clothing fashions of Lusaka to become instead markers of a highly contested and highly unequal landscape.

In January 2003, several elite Zambian business women were viciously attacked by marauding Christian fundamentalists in the Lusaka CBD for wearing mini-skirts and other clothes the attackers likened to prostitute clothing. Shocking as the attacks were, they were merely the latest in a series of skirmishes over the past decade or so. The anthropologist Karen Hansen (2000, pp. 216–17) has shown that the pattern in Lusaka of male violence toward young women they deem to be scantily clad motivated more than one-third of the young female high school students that wrote essays for her research to associate mini-skirts with rape. "Here in Zambia," one young woman wrote, "if you put on a mini and tight clothes, men can easily rip your clothes and you might be raped at the same time." Clothing symbolism, Hansen (2000, p. 229) suggests, is "at the center of a local battle" about culture that can be directly tied to the era of neoliberalism and good governance.

The contestations Hansen articulates via clothing choices in many ways parallel the conflicts that appear in James Ferguson's (1999) ethnographic account of the decline and fall of popular "expectations of modernity" on the Copperbelt at the dawn of the neoliberal era. Ferguson sees a kind of culture war between the lifestyles of those he terms "cosmopolitans" and "localists" in Copperbelt cities, but the categories transfer to Lusaka as easily as the tens of thousands of Copperbelters who have moved to the capital in the last decade (Bachmann 1991). Both styles are fundamentally urban, in Ferguson's view, but the latter style draws on "local" Zambian ideas of dress, music, housing, and custom in general, while the former embraces the styles of "the world out there" (Ferguson 1999, p. 215). Ferguson is at pains to distance his conceptualization from any simplistic traditional/modern dichotomous typology. Instead, he sees each style as an attempt to perform in a society in freefall, a manner of confronting the "noise" of modernity. Localists turn away from the "unintelligibility" of modern styles, while cosmopolitans seek the surprising, shocking, foreign, strange, or transgressive. One is reminded in reading Ferguson's account of cosmopolitans of an imaginary scenario from the writings of the Italian political philosopher Antonio Gramsci, where the peasants reply to an innkeeper's demand that they pay for the smell of his meat by clinking their coins. All is surface, nothing is whole or real, and rules are thrown out the window.

This dispute of localist versus cosmopolitan cultures has refracted in 73 directions in the past few years. This precise number is chosen because it is the number of ethnic groups officially recognized in the country. The past decade has seen a blossoming in local ceremonies for nearly every one of these, and struggles over royal succession, mismatches of political and cultural boundaries, and political party allegiances within these ceremonies have reached fever pitch. In July 2004, the country's leading opposition politician, Anderson Mazoka, was given the red carpet treatment at one such ceremony in his Southern Province home area, whereas Zambian President Levy Mwanawasa was unceremoniously shoved aside. Mwanawasa's New Deal (as opposed to New Culture?) administration claimed that it would fight the rise of "tribalism" in the country, but the national slogan of Kaunda's humanism, "One Zambia, One Nation," has looked in need of a replacement: One Zambia, 73 Nations (Sata 2004, p. 1; Maimbo 2004, p. 11; B. Phiri 2004, p. 1).

This extreme localism has yet to have a direct and obvious place in Lusaka itself, but, given the fact that all 73 Zambian nations are present in the city, one can see it subtly at work in specific ways. To do so, some reminders of Lusaka's cultural-historical geography are useful. Lusaka has always been a city fundamentally carved up by race, ethnicity, and, in the last half-century, party affiliation. These latter two especially are becoming more and more inseparable, but the links between them now are not without precedent. The new African housing areas of 1950s and early 1960s "introduced the classical principles of the colonial city to Lusaka" with their "clear separation from the European quarters" (Tait 1997, p. 204). Race and ethnicity were deployed to shape political party development at the very inception of most Lusaka compounds.

The population in unauthorized compounds expanded significantly in the 1950s and early 1960s (Tait 1997; Rakodi 1986b). Unauthorized locations and unsatisfactory employers' compounds were considered "slums of a deplorable public health standard" that the state should "take all possible steps to eliminate" (NAZ: LUDC 1/4/47). At the same time, however, such steps would have been expensive. With the failures of the public housing projects in the era, "an official toleration of squatter settlements" emerged among some officials (Tait 1997, p. 206). As Tait (1997, p. 206) argues, "the laissez faire attitude toward squatter development met the various changing needs of the growing colonial urban economy. It also reflected the inability of the colonial administration to ... control the African urbanization process." Some voices in the colonial regime found the means to justify their inaction by recourse to the cultural value of the unauthorized areas.

A 1957–58 study on African Housing in Lusaka Urban District compiled by the District Commissioner's office in collaboration with the Rhodes-Livingstone Institute offers an example of the ambivalence at the heart of the colonial state's dealings with the unauthorized compounds (NAZ: LUDC 1/4/24). On the one hand, the report is full of alarmist rhetoric. While acknowledging that compounds were "old established, and similar situations are a feature of many of the cities and larger towns of Central and Southern

Africa," the report's authors argued that the problem of compounds had "assumed serious proportions for the post-war 'boom development' in the Territory ha[d] led to an enormous increase in the Lusaka Africa population." That population was reported to have increased by 56 per cent from 1954–57, and the unauthorized locations grew by 138 per cent. No services existed at all in the unauthorized compounds in the Municipality, or in the peri-urban areas just outside it. The report found that 33,493 out of 75,210 Africans (or 44.5 per cent) lived in unauthorized compounds. These compounds could not "be properly and regularly policed." There were "no sanitary or refuse disposal services afforded to these locations and water [wa]s obtained from shallow open wells." Fully 53 per cent of the unauthorized households used the bush as a latrine, and another 30 per cent used a communal pit.

These and other conditions fostered what the report called the five political effects of the squatter problem. The first political effect was that "unauthorized locations corrupt[ed] the average African coming to Lusaka from a rural area." Secondly, the lack of stability was said to lead to "a lack of social conscience or sense of responsibility." Furthermore, "the physical disadvantages of life in these compounds afford[ed] opportunities for the agitator." That was especially worrisome because "the location of some of these locations adjacent to European residential areas d[id] not make for harmonious race relations." Finally, the District Commissioner's office warned of what it termed the "psychological effect" of compound living, in that a "man" there [wa]s "very apt to regard himself as opposed to the forces of law."

On the other hand, the report's authors could not help but recognize other aspects of the compounds as appealing. "It has been interesting to discover the extent to which the inhabitants of the unauthorized locations provide services for their fellows in the Municipal Locations." The District Commissioner's office stressed that it was mistaken to think of these people as criminals – "The great majority of them are very decent people." They had no alternatives, and moved to compounds for cheaper accommodations, nearer to work (NAZ: LUDC 1/4/24). The colonial state therefore directly and indirectly used as many of these "decent people" as it could to quell agitators, setting up channels of animosity within the compounds that eventually came out in the wash as party affiliations.

Party politics from the late colonial era onward further carved the city into wedges and sectors. Lusaka's townships and compounds lined up behind the African National Congress or the UNIP, and there was seldom much toleration of the other party in the strongholds of either one. For instance, Hansen (1997, pp. 64–65) has shown that the squatter compound of Kalingalinga had very strong ties to the ANC that did not really disappear with the declaration of a one party state. The model site-and-service scheme that the UNIP government developed in the late 1960s and early 1970s just east of Kalingalinga, in Mtendere, became filled with UNIP loyalists who had previously resided, uncomfortably, in Kalingalinga. The actual parties to which the two compounds generally gave their loyalties changed between the 1991, 1996 and 2001 elections, but party differences between the two compounds remained in the middle of the first decade of the 21st century. By

2004, Mtendere was the heartland of support for Michael Sata's Patriotic Front, where Sata might have had trouble drawing a crowd in Kalingalinga. Support for Sata in Mtendere did not come out of nowhere – he was the Governor of Lusaka under the UNIP Second Republic government that had built many of the amenities and infrastructural advantages that Mtendere still to some degree enjoyed.

In late November and early December 2002, the Lusaka City Council began a campaign of demolitions in squatter compounds throughout the city, including Kalingalinga, Ng'ombe, and Kamanga. This was actually a follow-up program to one initiated in 1999 by the previous, MMD-led Council (*Insight* 1999b). With the somewhat reluctant assistance of the police, the Council demolished thousands of homes of varying quality and stages of completion in nine different compounds. The Mayor claimed that the Council had "targeted structures that were under construction and not yet occupied," but at least 700 families in one of the compounds alone, Ng'ombe, were made homeless and became "refugees" in the neighboring compound of Mandevu. The rains began just as the homes went down, and the "squatters [we]re living in squalid conditions in tents" (*Times of Zambia* 2002a, p. 3). Diarrhea, dysentery, and cholera ensued. The residents of another compound, Kalikiliki – the unauthorized overspill area from Mtendere – actually fought back, and torched the police bulldozer in the process; but when the police returned, they were firing live bullets, and two teenagers were shot and killed. The Mayor "deeply regretted the loss of life" in what was still touted as a "well intended demolition exercise" (Habeenzu 2002, p. 7). His Deputy said "the demolition exercise that LCC was carrying out was not aimed at making people destitute but [at] bring[ing] sanity [to] the city" (*Times of Zambia* 2002b, p. 1). A similar "sanity" lay behind the Mayor's "Keep Lusaka Clean" campaign, through the Keep Lusaka Clean Trust Fund, which "aimed at ensuring that vendors are removed from the streets once and for all" (*Times of Zambia* 2003b, p. 2), since street vendors had become such an eyesore for the city's business elite.

The roots of the conflicts over these demolitions are complex, and they differ a bit in each particular compound. What is consistent in each case is that the plots on which the demolished structures were constructed had originally been sold to the newly displaced people by local MMD officials in the lead-up to the 2001 election campaign. With the demolition orders of a new council completely dominated by non-MMD councilors, local MMD leaders in the wards felt that they would have the power of the MMD central government behind them. That was not enough to stop the bulldozers, but it did come into play in the end, albeit somewhat discursively.

Kalikiliki had long been a difficult neighborhood for Mtendere residents. Several spoke to me of their sense of fear in passing through it. Mtendere, after all, had been built in the Kaunda era for a largely UNIP base, and though it certainly did not stay a UNIP neighborhood, it has never been a comfortably MMD one. Most Kalikiliki residents are newcomers, northerners, and MMD supporters. This regional division finds a home in the stories of the demolitions in the other peri-urban compounds, as well.

Although Kalikiliki's demolitions and the violent resistance to them resulted in deaths, it was the displacements in Ng'ombe that lasted longest and had the deepest repercussions. Fully 700 Ng'ombe families were still living in UN refugee blue tents in Mandevu in July 2004. Even though they were given plots on which to build, they had few resources with which to do so in most cases. While churches – and particularly charismatic, pentecostal churches – have provided central social bases for the development of communities in compounds during the crisis years of the last few decades, in the over-whelmingly Roman Catholic streets of Ng'ombe, a counter-dynamic appeared instead. For many refugees, their dispossession shook the core of their religious faith, for a very specific reason. The land that MMD cadres had sold them in Ng'ombe was claimed by the Archdiocese of Roma Township. "Some people threw off their crosses and rosaries after this, saying we won't be in this church any more," one of the displaced persons told me. It was not their church, but the MMD-led central government that they then looked to for justice.

In some strange way, these MMD cadres were not wrong in anticipating the central government's reactions; they were just a bit too late to save their homes. Legally, land can only be allocated through the central government channels, and the City Council was certainly within the law to order the demolitions. Yet the MMD-run central government's District Administrator for Lusaka made a major case of the refugees from Ng'ombe, providing them with plots at the expense of the many thousands of people who fail to obtain land legally (*Times of Zambia* 2003a). In a none-too-transparent embarrassment of the City Council, the central government-controlled Times of Zambia newspaper and Zambia National Broadcasting Corporation television and radio outlets gave prominent placements to the plight of the Ng'ombe refugees for weeks on end.

Most of those displaced were recent migrants from the MMD stronghold of the Copperbelt. Yet the politics of cultural difference that played out in this case ended up being more complex than a simple Bemba/Nyanja, Copperbelt/Lusaka, or northerner/easterner rivalry. Significant differences across party and regional lines and within alliances came out in public regarding the utility and timing of the demolitions, as well as inconsistencies in the removal policies. The way Zambia's vibrant press and popular opinions jumped on both stories marked a dramatic change. One editorialist critiqued the District Adminis-trator for "posturing as the godfather of the displaced people from Ng'ombe." He questioned the "incessant television coverage" of the "victims," arguing that "yes, the story must be told, but that should be within the context of all those people who live in deplorable conditions in squatter compounds ... It would only make sense if this sad situation was used as a wakeup call to embark on a vigorous campaign to change the living conditions of all our poor people who live in squalor in all these compounds" (Komakoma 2003, p. 8).

People in Ng'ombe and Kamanga offered rather clear-eyed assessments like this: "the central government is so slow in processing applications for plots. The people in charge of allocations take money in order to issue approvals so it is cheaper and shorter and less of a hassle to simply squat. Then council sits back until the construction reaches the roof and then they say they want to demolish. Come on. I mean, why can't they just issue land to the people"

(Mumbi 2003). The open multi-vocality of public space in early 21st century Lusaka belies any simplistic dismissal of the time as a mere throwback to the party politics of the past or divisions of late colonial times.

Still, the demolitions could not have occurred at a worse time for the SLP's community-based enterprises in the compounds. What kind of a signal does it send about the sincerity of efforts to generate a "new culture" of business-like, democratic, and decentralized planning when the follow-up procedure is a Caterpillar on full throttle? Even if the CBEs could see themselves as independent of the City Council, to the residents of the compounds they could not be seen this way at all. A council that bulldozes homes built on land obtained in almost exactly the same way that most residents of the compounds got their own land and housing is unlikely to be a council that generates a participatory and democratic system of solid waste management. The CBEs can't shake the Shitty Council.

Lusaka's politics of cultural difference seldom take on the ominous and even global tones of those in Dar es Salaam and Zanzibar. Yet disputes between localist and new-culture cosmopolitan identities, disconnections between "shacks and mansions" (Knauder 1982), clashes between party cadres whose loyalties to the party overlap substantially with ethnicity and region, and conflicts between communities and the Council all played a role in under-mining any legitimacy the SLP might otherwise have earned for itself in its solid waste initiatives in the compounds. The contemporary trajectories of these fracture lines in urban Zambia do not bode well for the future possibilities for progressive grassroots planning initiatives.

Conclusion

There is no question that a new "nexus of truly developmental state-society arrangements" depends upon "greater citizen participation in decision-making" (Cheru 2002, p. xiv). Lusaka, like other African cities, clearly needs "more competent, accountable city government ... to lessen localized environmental burdens" (McGranahan and Satterthwaite 2002, p. 221). The donors for various SLP initiatives at least said that they intended their moneys to be used for creating greater citizen participation and competent, accountable city government. But the SLP can probably be placed in the "credibility gap" that I discussed in chapter 2, which Lubelski and Carmen (1999, p. 110) identify with the Local Agenda 21 network of urban sustainability planning. Behind its rhetoric of participatory, democratized planning for and with the marginalized urban poor, international development priorities for privatization and donor-conceived definitions of participation actually predominate. As Lubelski and Carmen (1999, p. 116) put it, "truly sustainable development cannot be planned as a grand scheme and imposed on people, rather it must be endogenous, arising from people's own specific experience and situations."

A program that seeks to decentralize control and devolve authority over urban service provision in the absence of "complementarity between government and citizens" will in all likelihood result instead in "the disempowerment of local authorities and local societies alike" (Beall et al. 2002, pp. 22 and 65). And this seems to have been the case in Lusaka. To ordinary citizens with a "deep distrust of the state and a perception of its institutions as irrelevant to everyday life" (Abrahamsen 2000, p. 55), no institutions earn greater distrust or are viewed as more irrelevant than those of local authorities and city councils, given their exclusionary histories.

The SLP seems to have done little in the long run for sustaining the community based enterprises it fostered. The thrust of their work has now been usurped by larger private firms bidding to remove trash from compounds, but only if there is real money to be made there. The SLP seems therefore to have been at best a story of an incomplete empowerment of civil society, despite its having created and trained the CBEs in the first place. The exclusionary governance at the heart of the larger political and economic framework in which Lusaka operates left it at the mercy of external donors, whose own internal changes dictated its finances. The social democratic northern European donors might claim to have a stronger commitment to a government role in urban service provision. But in Lusaka's SLP, they too have failed to sustain local, grassroots initiatives aimed at trash removal.

Lusaka's contrasting spaces of daily life – production, social reproduction, and consumption – cannot help but prompt questions of where, if ever, rhetorical and institutional versions of sustainable development intersect with environmental justice. It is hard to see how institutional developments like the SLP matter to vague questions of sustainability, governance, or neoliberalism that paves the way for glitzy shopping malls that 75 per cent of the population can't set foot in when the environmental injustice of the city is so painfully, physically reinforced every day. Preventable disease and death haunt compounds in even more gruesome ways than they ever have before – it is a truly sad commentary on the last fifteen years that one of the most lucrative informal sector businesses in a compound is now casket-making. To the extent that the near-total failures of solid waste management in compounds are involved in this haunting – and that extent is not inconsiderable, given the localized roles of garbage in flooding, air and water pollution, or pest or disease vector incubation and infestation – garbage is arguably one of the fundamental political ecological issues of the years of money. In part because of the SLP, Lusaka compounds now have a wide range of dynamic citizens engaged in NGO-led developments that link them with progressive global forces. New churches spring to life with new ministries by the month. Yet neoliberalism, good governance, sustainable development, and the politics of cultural difference have set up shop and show no signs of leaving town. The garbage is still there. And the casket makers continue to thrive.

Chapter 6

Conclusion

Cities are places from which people potentially can change many things. ... But change is something that demands platforms and resources. If communities are always defending themselves against the city, how does the city become an instrument for change?

(Simone 2004, p. 213)

Introduction

I began this research project with a genuine enthusiasm for the ideals and proscriptions of the Sustainable Cities Program in African cities. I wanted to construct – and thought I saw the makings of – a counter-narrative to the prevailing sense that "African cities don't work" so common to both popular and scholarly understandings in the west (Simone 2004, p. 1). On paper, the participatory, grassroots Environmental Planning and Management framework seemed to me at the very least a vast improvement on past planning and at most a revolutionary idea. Having had more than two decades of experience living, studying, and researching in eastern and southern African cities, it seemed to me inescapable that environmental problems like those stemming from inadequate solid waste management needed to be addressed. The EPM approach seemed to me a practical way of balancing community participation with a strengthening of local states.

Here and there, I can see elements of what I thought or hoped I would find in the three case study cities I've studied. In people like Peter Tembo, Ruth Mundia, Mzamose Mbewe, Bimkubwa Ali, or Kasu Bachu, I found ordinary working class Africans trying to take charge of the solid waste crisis around them in their poor neighborhoods. Government officials and working group leaders like Sheha Juma, Makame Muhajir, Litumelo Mate, Martin Kitilla, Daniel Nkhuwa, or Joash Nyitambe suggested to me that what AbdouMaliq Simone (2004, p. 1) calls a "more generous point of view" might indeed be warranted, one that sees these cities as "works in progress." On the surface, Dar es Salaam is a somewhat cleaner city than it was a decade ago, and both Dar and Zanzibar have programs in place to recycle some of their solid waste. All three cities at least have a few relatively new avenues for popular agitation over policy, and some broken shards of local electoral democracy.

Ultimately, though, I completed the research with a jaundiced eye on the realities of the program's accomplishments or even possibilities. In this conclusion, I want to address four reasons for my doubts, connecting the case

139

studies to broader claims. It is no surprise by this point in the book that these reasons have to do with neoliberalism, sustainable development, good governance, and the politics of cultural difference. I discuss each in turn in the next four brief sub-sections.

Neoliberalism or ubinafsishaji*: The Years of the Rule of Money and Garbage*

The last fifteen years have brought vast changes to the Tanzanian and Zambian economies. While the IFIs and donors appear moderately pleased with the pace of reform adoption in Tanzania, and their views of Zanzibar and Zambia are more mixed, assessments from the majority of urban residents have been much more critical of this era of transformation. The labels that urban Tanzanians, Zanzibaris, and Zambians give to the neoliberal era change, but the general intents behind the terminology are similar. This is a time of privatization and the rule of money, where policies generally work against the best interests of most of the people.

Neoliberalism's main impacts on the lives of urban Tanzanians and Zambians are economic: unemployment and under-employment have never been higher. But the progress of the SAP/PRSP/HIPC agenda also has grave political implications. "Reductions in the size of the civil service, subsidy cuts, privatization, and deregulation under structural adjustment programs have eroded the state's capacity to intervene effectively in the economy and in social policy" across Africa (Simone 2004, p. 186). In some cases, the trend runs toward state collapse and anarchy. In other cases, like Tanzania and Zambia, neoliberalism just leaves states largely incapable of exercising autonomy in policy development. This is also the case for urban policy. Even the role of pilot program for the world earned by Dar es Salaam in the Sustainable Cities Program, after all, came about when the United Nations said it would only support a new plan for the city if the city followed its (neoliberal) plan. The starved and warped local governments of Zanzibar and Lusaka have proven even more incapacitated for the program's expectations.

Neoliberalism's economic consequences for each of the three cities have been severe. First, poverty has become more – not less – widespread and more consequential in each city. Second, precarious urban local governments have lost financial capacity and autonomy. As a result, each is in varying stages of handing over authority to private firms for solid waste management that uniformly under-serve the poor majority. Even in the one city with a significant improvement in its rate of landfill deposition during the era (Dar es Salaam), that improvement has overwhelmingly impacted the city center and wealthiest parts of the city. Those who can pay for waste services sometimes receive them now, even in Zanzibar or Lusaka. Virtually everyone else is out of luck, choking in their own burning waste. The highly touted employment generation dimensions of privatization have not played out to the advantage of very many people. In the widest example of Dar es Salaam, the number of private firms and number of employees have contracted considerably, rather than expanding. The wages paid to the handful of employees of experimental community-based enterprise schemes in Zanzibar and Lusaka are minuscule, and even the owner-operators of these schemes have lost money. In all three cases, the new regimes of solid waste management seem highly uneven in the distribution of any benefits from either privatization or decentralization.

Beyond Sustainability Rhetoric: Environment and Development in Urban Africa

Neoliberalism is also much more than an economic philosophy. Neoliberalism has had serious environmental consequences in the case study cities. The Other Sides of these cities, like those of most African cities, still don't get their garbage picked up. Large numbers of poor people suffer and die from easily preventable diseases that stem from this failure. The Sustainable Cities Program set out to change this crisis dynamic from within, by building community capacity for solid waste management. In a few cities, for a limited period of time, rates of collection and deposition record temporary increases. Yet these increases, such as those credited to the Sustainable Dar es Salaam Project, did not in the slightest rest on the progressive, bottom-up, deliberative, or argumentative planning that is supposed to be the bedrock of the sustainability component of the EPM framework. Arguably, the case with the greatest nods toward that sort of progressive planning – that of the very argumentative compounds of Lusaka – has seen a complete stagnation of its collection rates and a total abandonment of support for its progressive dimensions from donors and government. Zanzibar had a higher rate of solid waste collection and deposition before the ZSP began. One might be led by this to conclude that urban dictatorships collect more garbage but create a less emancipatory or inclusive city. But that is a facile conclusion. Things are worse than that, because in any case all of the cities seem to be moving further away from inclusive governance and further away from effective solid waste management at the same time.

"New Geographies of Governmentality" in African Cities?

Decentralized, localized, and community-based development rhetoric has achieved sacrosanct status in the development universe. One route toward that inscription of the local as the better angel of our nature comes through the articulation of a post-development future where development decisions are "decentralized, community-based, participatory, indigenous and autonomous" (Watts 2003, p. 9; Escobar 1995). Sanguine paeans to communities creating "new geographies of governmentality" produce a sub-text of nearly utopian idealism in this literature (Appadurai 2001). As Watts (2004, p. 197) shows, though, this obsession with decentralized communitarian governance is "one of the defining articulations of neoliberal rule," where "communities arise from the ashes of state withdrawal and speak the name of civic renewal." To be sure, it is rather easy to point out "the failure of the centralized African state" to produce development benefits for a broad majority of the population of countries like Tanzania and Zambia (Wunsch and Olowu 1990, p. 1). It is quite another thing to champion state withdrawal and community autonomy in the context of neoliberal triumphalism. In that context, we are still far from any democratization of governance from below.

We might say that new social networks have arisen out of the formation of the various community-based groups in these programs that link, in various ways, with state agencies and private capital to attempt anew to collect the waste of poor neighborhoods. Yet these networks themselves are structured in a governance ethos that is, ultimately, authoritarian. The rhetoric does not match the reality for the specific reason that it is not supposed to. EPM is much more about talking about liberation or participation than it is actually liberatory or participatory. The actual "conduct of conduct" that comprises Michel Foucault's idea of "governmentality" remains exclusionary (Watts 2003).

In each of the case study cities, the last twenty years or so have seen a nearly constant stream of shifts in state structures to stay on top of city matters. Superficially, most of these shifts appear to be decentralizing decision-making further and further down the hierarchies of the state apparatus even as electoral politics appear to put more and more power in ordinary people's hands. The institutional landscapes are different in each of the cases, and the degree of fractiousness to politics varies. But ultimately, when we start to ask who "owns" the projects and policies enacted within the SCP rubric in Dar es Salaam, Zanzibar, and Lusaka, our answers move away from the supposed downward restructuring of the relationship between people in poor neighborhoods and the government. The ideas from below hardly matter at all to policy development or outcomes, particularly where the working groups structures are extraordinarily top-heavy with government personnel and where expatriate consultants or donors drive the program from beginning to end.

"New Culture" and the Politics of Cultural Difference

Culture and geography – and cultural geography – seem to be more prominent as explanatory themes in the literature of development than at any time since the demise of environmental determinism (Sachs 2001; Herbst 2000; Huntington and Harrison 2000). To most geographers, the deployment of these themes appears to be a simplistic throwback with a "pungent, late-Victorian imperial odor" (Watts 2003, p. 12). The mainstream academic attentiveness to what it sees as cultural geography plays out in the policy documents of donors like Danida, which highlighted the "power of culture" in its 2002 realignment of policy with staunch neoliberalism. But nuances, complexities, subtleties, and historical contexts in battles over identity and place are swept aside, or cultural differences are re-deployed in cookie-cutter fashion to perpetuate unequal systems of service provision. We either face assumptions of pious harmony or assumptions of insurmountable, structured differences.

The relations of power that have underlain the EPM framework as it has been employed in these three cities in the interest of improved solid waste management are fraught with inequalities and injustices that largely prefigure the framework's failure. In the end, most people thus continue to be left to themselves and their own devices, for dealing with waste and for dealing with life. "Heightened levels of instability" brought on by global processes, however, have steadily eroded the capacity of locally-formed social networks of reciprocity for dealing with, in some idealized communal form, the large gaps in basic provision that have widened with this instability (Simone 2004, p. 225). The alleged stakeholders of the framework have no stakes to claim. The "new cultures" of these three cities remain in mourning mired in a crisis of meaning.

(Post)Script:
Postcolonial Geography
and Urban Policy

The brilliant 1981 film, *Black and White in Color* (which won the Academy Award for Best Foreign Language Film) is a telling one for geographers. The film depicts World War One as it was fought in two villages on the border of the German colony of Togo and the French territory of Dahomey (Benin). In the movie, the white elites on each side conduct a hideous and farcical campaign that pits ragtag armies they forcibly conscript from amongst the Africans they think they are ruling. The first battle goes horribly awry, even though the whites on both sides had encamped for picnics expecting to be entertained watching the struggle between their forces. Each side's army is then trained and made efficient by the schoolteachers of the two respective border villages. When the two sides find out the war is over – after it has already ended in Europe – the colonizers call the battles off. The two schoolteachers are seen walking off into the sunset in the final scene, arm in arm. "You know, I was a geographer before the war," the German says.

The history of modern human geography in Africa is not a long one or a pretty one. As a disciplinary practice, as *Black and White in Color* vividly (pun intended) demonstrates, it is thoroughly bound up with the history of European imperialism and colonialism. The "discipline of geography," Jenny Robinson (2003, p. 277) writes, "has a past littered with the skeletons of murderous neglect and encounter," and sub-Saharan Africa was surely a central killing field. The first professional geographers operated as explorers, or forerunners, for imperial conquests in the era that Felix Driver (2001) portrays – borrowing from Joseph Conrad – as "geography militant." These early geographers, as Watts (1993b, p. 174) put it, "helped make Africa 'Dark'" even while claiming to help show the way for "the 'Darkness' to be lifted." With the establishment of colonial rule, the practitioners of the trade settled into a role I like to think of as geography *mundane* – they worked as the surveying and inventory-making agents of colonial regimes. Occasionally, the district officers of empires actually did the work of geographers, as in the British program of the 1930s that requested human geographies of each region of each colony to be written by the district officers (Stone 1978/9, p. 31).

Few geographers working outside of the colonial service constructed cultural studies of the continent, with the vast exception of South Africa. When such studies did appear, they generally followed the trends of the times and stayed

144

well within the political boundaries of colonial apologism. For many years, a cultural-ecological regional-inventory approach held sway; land use patterns were mapped, but without much conceptual light on the political-economic processes behind the patterns (Light 1941; Stamp and Morgan 1953; Hance 1964). Only occasionally did critical or progressive voices appear in this time of geography mundane (as in Buchanon and Pugh 1955).

From the 1950s onward, geographical research in Africa became virtually inseparable from what came to be called development. The 1960s brought modernist modelers and spatial scientists of modernization to the continent, where reams of maps were made to show the spread of telephone lines and the waves of progress in conventional terms (Soja 1968). Despite the general enthusiasm for independence such geographers showed, critical cultural analysis was decidedly not their forte, at least at that juncture in their careers. Political geography, too, was almost completely absent, and urban geography studies played second fiddle (Watts 1993b; Mabogunje 1968; de Blij 1963). Most modernization geographers from the West rejected their own approaches and findings within a dozen years, despite the influences their works continued to have on the first post-independence generation of African geographers (Riddell 1981; Darkoh 1994).

The stall of modernization and development engines across the continent brought a more radical edge to the field by the 1970s, with political economy becoming the bedrock for critique in "development" geography. Watts (1993b, p. 182) saw this as a "watershed in African studies more generally," since Africa-based analyses – really for the first and only time – became somewhat central to debates in a variety of social science disciplines. Critical development geography crested in the wake of this watershed, though, during the 1980s. Critical development geography ultimately was put in a box – one that used to be labeled, in turn, colonial geography, tropical geography, or regional geography – and in its utility for the field as a whole it was placed on the equivalent of a high shelf far away from the center of the room (Robinson 2003, p. 278; Power and Sidaway 2004).

Postcolonial studies, then, came into geographical study of Africa at yet another impasse, as the critiques of radical development geography seemed to make little headway in transforming patterns on the ground (or in the conference halls of the western academy). Development itself came to be widely recognized as "a particular vision that is neither benign nor innocent" (Power and Sidaway 2004, p. 594). Postcolonial studies offered a "set of theoretical perspectives" or conceptual strategies aimed at "hearing or recovering the experiences of the colonized" (Sidaway 2000, pp. 591 and 594). Postcolonial theory aimed at getting past the post, at contesting the lingering and debilitating modes of thought and action that comprise postcolonial conditions. Postcolonialism didn't just come after colonialism temporally; it came *after it* with a vengeance to contest its legacies (Loomba 1998, p. 12).

Some in the field readily decried the "descent into discourse" and retreat from political engagement that the engagement with postcolonial studies seemed to imply (Barnett 1997, p. 137). David Simon (1998, p. 221), for instance, expressed frustration toward the postcolonial studies literature for what he saw as its disconnection from actual studies of development, beyond an idealizing of local level social movements and non-governmental organizations by some thinkers associated with it. As part and parcel of a broader "retreat into theoreticism" in cultural geography, postcolonial studies geography severely limited "the purchase of its intellectual products" (Robinson 2003, p. 275).

I still believe that there is much work to do, though, utilizing some semblance of a postcolonial lens to interrogate contemporary development planning such as the urban management schemes I have studied in this book. What I mean by a postcolonial lens is one that takes the discursive tactics and legacies of colonialism seriously, but without losing sight of what Derek Gregory (2004) calls the "colonial present", with its very material reality of domination and inequality. Let us not get lost in the past or in discourse, for postcolonial geographies are contemporary material realities of injustice (McEwan 2003b). I see this book as part of a broader attempt to create a decolonized geography, beyond colonialist ways of thinking and doing academic analyses of African urban development. My idea in writing this book has been that we can re-materialize cultural geographical inquiry in Africa, without losing the conceptual strengths of postcolonial studies. Materialism, not in a corollogical sense, and not in a strictly determinist sense that misses complex interplays of the cultural arena, can allow us to see culture in motion, in process, as constructed and lived and contested landscapes, not dead collections of traits.

Postcolonial geographies can aim at articulating and contextualizing the poisonous hangover of colonialism, in both conceptual and practical ways. Conceptually, postcolonial geographies need to be decentered from the beginning, not ignoring but also not dependent on western modes of thought or trendy social theories hip in Paris, London, or Los Angeles, and instead building from the knowledge bases of Dar es Salaam, Zanzibar, or Lusaka. In this conceptual decentering, of course, the first thing we confront is the reality of the interpenetration of ideas, the hybridity of both imaginary schools of thought on the world ("western" and "African"), from far back on the timeline until now. Rather than being immobilized by this recognition, though, we must use it to balance and truly globalize the exchange of ideas in geographic thought, in development studies, in urban theory. In geographer Brenda Yeoh's (2001, p. 464) words, "the prospects for a postcolonial urban geography must be tied to the larger enterprise of constructing and elaborating alternative postcolonial geographical traditions, for without these new lenses how can we hope to grasp the swirl of sensibilities and groundswell of politics emerging in the postcolonial urban world?"

This book contains a modest foray into mapping some of this alternative thinking, in particular in my reliance on African literatures of urban studies and development geography – works by Africans, often works produced in Africa – in combination with African insights that emerged in interviews, and with some Euro-American social science. Probably the broadest conclusion that could be reached from a reading of these works or interview transcripts for me, though, was that postcolonial geographies could not be only about theories and thoughts. Postcolonial conditions are real and need to be confronted, for as Yeoh again (2001, p. 459) writes, "it is often in the buzz of the streets and the thick of urban encounters between individuals and groups that the postcolonial is enacted and lived, often in contestatory terms." One central element of a decolonized vision of the discipline of geography, or of urban studies, ought to be one that is infused with activism and inter-disciplinary, materially-engaged inter-cultural research. Probably more than anything else, this means people like me – white male western outsiders – listening to and working from the beginning with colleagues and partners in African universities and societies.

The 800-pound gorilla in the room when this kind of collaborative energy begins to spark is the glaring structural inequality of the global economy, along with its twin, the highly unequal "geopolitics of knowledge" that grants westerners like me significant privileges in disseminating research findings (Robinson 2003, p. 274). Most of Africa is, ironically, "becoming increasingly peripheral in economic terms" (Briggs and Yeoah 2001, p. 18) as contemporary globalization proceeds. The world is turning away from Africa. In many countries, both urban and rural poverty grows deeper just as political crises, health crises, and environmental crises mount, yet western aid donors, investors, and even scholars in effect seem to have, like Elvis, left the building. We need to bring the world and Africa back into conversation with each other, but on new terms, ones that take a vast range of African perspectives seriously from the start, on par with everyone else's ideas.

There are several talking points that must appear in such a conversation. The first is, of course, inequality. In societies living under postcolonial conditions, cultural studies more generally, but particularly those driven by Euro-American trends in theory, can appear as an alien imposition, as an imperialism that by any other name would still smell as rotten (Karp 1997). "The shift toward cultural explanations and concerns with discourse and representation," Cheryl McEwan (2003a, p. 410) writes, has for one instance "been ridiculed by many activists (primarily in the south ...) as elitist and removed from reality." Writing particularly about feminism but in a manner applicable to other approaches within human-cultural geography, she notes part of the reason why: "theoretical preoccupations are not easily translated into direct politics" (McEwan 2003a, p. 414). Many central subjects of inquiry within Anglo-American cultural geography can readily appear as "side issues" (McEwan 2003a, p. 414) to African scholars or activists faced with real and present crises of violence, disease, poverty, or injustice. Many African and Africa-oriented geographers undoubtedly worry, with Don Moore (1997, p. 103), about "turning ... livelihood struggles into fodder for conceptual refinement [where] ... the politics of place becomes subordinated to academic sites/cites of struggle." Yet the politics of place can be so intertwined with those real and present points of crises as to demand an understanding of cultural experience, as I have tried to demonstrate in this book.

A second talking point for a decolonized, *post*colonial geography is simply to bring African urban studies into the global conversation on cities in and around the discipline. Most western-world urban studies scholars have yet to really wrestle with the diverse materiality of urban space and its contestation in Africa. Many debates in urban geography and urban studies have left Africa "off the map" (Robinson 2002; Sidaway 2000). For one particularly glaring example of the consequences of this lacuna, the excellent edited volume on *Wounded Cities: Destruction and Reconstruction in a Globalizing World* (Schneider and Susser 2003, p. 3) had no chapter on urban Africa – despite the editors' apparent attempts to overcome the "unfortunate" gap – when it would seem painfully obvious to me that the continent would have to lead the world in defining the very nature of woundedness in cities and among city-dwellers. More generally, this sort of failed engagement is a crucial flaw because, as Mabin (2001, p. 181 and 183) puts it, "understanding contemporary global urbanism requires more consideration of social circumstances in cities beyond the continents of Europe and North America," the cities of which "decreasingly represent the urban world of the twenty-first century."

The immediacy of crises and propulsive energy of growth that surround one in the conduct of research of almost any kind in African cities literally begs one "to engage outwards, with a world which lies beyond our own internalized conversations" (Massey 2001, p. 12) in academia. In part, this means engaging the "policy turn" – as this is manifested, for me in this book, in the Sustainable Cities Program. By engaging policy, I do not mean that what we do must fit conventional confines of policy relevance (Henry et al. 2001). Instead, what I mean is a direct, material engagement with "what is happening politically" as a result of policy rather than "apolitical [policy analysis] draped in the terminology of 'good governance'" (Freund 2001, p. 718). For urban, political, and cultural geography, as well as for urban studies, African cities clearly provide "outward" spaces beyond the internalized conversations. "African cities have been platforms of mediation" for the central issues of our age in the world (Simone 2004, p. 18). My analysis of the ongoing restructuring of urban environmental governance in Africa – largely framed by African perspectives on neoliberalism, sustainable development, good governance, and the politics of cultural difference – is meant as a contribution to an understanding of the ongoing transformations of development in the world, and not in an isolated, or exceptional, Africa.

I see the book as an example of postcolonial cultural geography not simply because I am dealing with questions of the politics of cultural difference in societies that were once British colonies and are no longer. The credo that Abrahamsen (2003, p. 210) suggests that postcolonial studies can adopt from Michel Foucault in analyses of African culture and politics is appropriate here. In one interview, Foucault (1997, p. 256) said of his work, "my point is not that everything is bad, but that everything is dangerous. ... If everything is dangerous, then we always have something to do. My position leads not to apathy but to hyper- and pessimistic activism." Much of what appears in this book can be read as being rather pessimistic, to be sure. Yet it is also, I hope, suggestive of the "realm of the possible" for a new kind of "insurgence" (Holston 1995, p. 48) that not only aids the structural adjustment of the discipline of geography but reshapes the conditions of existence for people like Kasu Bachu or Ruth Mundia in cities like Dar es Salaam, Zanzibar, or Lusaka.

Bibliography

Abrahamsen, Rita (2000), *Disciplining Democracy: Development Discourse and Good Governance in Africa*, London: Zed Books.

Abrahamsen, Rita (2003), 'African studies and the postcolonial challenge', *African Affairs*, **102**, 189–210.

Adam, Haji (2003), Author interview with Zanzibar Director of Urban Planning and Surveys Haji Adam, August 7, 2003.

Afro-Shirazi Party [ASP] (1974), *Maendeleo ya Mapinduzi ya Afro-Shirazi Party, 1964–1974* [Development of the Revolution of the ASP, 1964–1974], Zanzibar: Afro-Shirazi Party.

Agyemang, O., B. Chirwa, and M. Muya (1997), *An Environmental Profile of the Greater Lusaka Area: Managing the Sustainable Growth of Lusaka*, Lusaka: Lusaka City Council.

Ahluwalia, Pal (2001), *Politics and Post-Colonial Theory: African Inflections*, London: Routledge.

Aina, Tade Akin (1997), 'The state and civil society: politics, government, and social organization in African cities', in Rakodi, Carole (ed.), *The Urban Challenge in Africa: Growth and Management of its Large Cities*, Tokyo: United Nations University Press, pp. 411–46.

Aina, Tade Akin, F. Etta, and C. Obi (1994), 'The search for sustainable urban development of Metropolitan Lagos, Nigeria', *Third World Planning Review*, **16** (2), 201–219.

Agyemang, O., B. Chirwa, and M. Muya (1997), *An Environmental Profile of the Greater Lusaka Area: Managing the Sustainable Growth of Lusaka*, Lusaka: Lusaka City Council.

Andreasen, Jorgen (2001), 'The legacy of mobilisation from above: participation in a Zanzibar neighborhood', in Tostensen, Arne, Inge Tvedten, and Mariken Vaa (eds.), *Associational Life in African Cities: Popular Responses to the Urban Crisis*, Uppsala, Sweden: Nordiska Afrikainstitutet, pp. 263–81.

An-Nuur (2004), 'Wapiga kura wa Keenja wadai kusahauliwa' ['Those who voted for Keenja claim to be forgotten'], *An-Nuur*, 21 May 2004, p. 4.

Appadurai, Arjun (2001), 'Deep democracy: urban governmentality and the horizon of politics', *Environment and Urbanization*, **13** (2), 23–43.

Armstrong, Allen (1987), 'Master plans for Dar es Salaam, Tanzania', *Habitat International*, **11** (2), 133–146.

Armstrong, Allen (1992), 'Dar residents refuse refuse', *Town & Country Planning*, **61** (10), 282–84.

Arnold, Nathalie, Bruce McKim, and Ben Rawlence (2002), 'The bullets were raining: the 2001 attack on peaceful demonstrators in Zanzibar', Human Rights Watch report 14 (3a), New York.

Asquith, Julian (1970), Former Zanzibar Protectorate official Julian Asquith, interviewed by I. Phillips, transcript in Mss Brit Emp s 375, Rhodes House Library, Oxford.

Bachmann, J. (1991), 'Reading the urban text: the legibility of modern values in the squatter settlements of Lusaka, Zambia', unpublished MA thesis, University of California, Berkeley.

Bales, Kevin (1999), *Disposable People: New Slavery in the Global Economy*, Berkeley: University of California Press.

Bana, Benson A. (1995), 'Human resource management in Tanzanian local government institutions: the case of Dar es Salaam City Council', unpublished MA thesis, University of Dar es Salaam.

Banda, Agatha Nkhaiko (2002), 'Solid waste management in Kamanga, Lusaka – Zambia', unpublished BA research project report, University of Zambia.

Banda, Azwell (2004), 'Are we political retards?', *Zambian Post*, July 11, 2004, p. 10.

Banda, Japhet (2004), 'Is HIPC the answer to Zambia's woes?', *Times of Zambia*, July 5, 2004, p. 5.

Banda, Naomi (2002), Personal communication with Kalikiliki resident Naomi Banda, December 17, 2002.

Banyikwa, W. (1989), 'Effects of insensitivity in planning land for urban development in Tanzania: the case of Dar es Salaam', *Journal of Eastern African Research and Development*, **19**, 83–94.

Barkan, Joel (1994), 'Divergence and convergence in Kenya and Tanzania: pressures for reform', in Barkan, Joel (ed.), *Beyond Capitalism vs. Socialism in Kenya and Tanzania*, Boulder, Colorado: Lynne Rienner, pp. 1–45.

Barnett, Clive (1997), 'Sing along with the common people: politics, postcolonialism, and other figures', *Environment and Planning D: Society and Space*, **15**, 137–54.

Barrow, C.J. (1995), 'Sustainable development: concept, value, and practice', *Third World Planning Review*, **17**, 369–386.

Bartone, Carl (2001), 'The role of the private sector in municipal solid waste service delivery in developing countries: keys to success', in Freire, M., and R. Stren (eds.), *The Challenge of Urban Government: Policies and Practices*, Washington, DC: The World Bank, pp. 215–23.

Batley, Richard (1993), 'Political control of urban planning and management', in Devas, Nick, and Carole Rakodi (eds.), *Managing Fast Growing Cities*, Harlow: Longman Scientific and Technical Publications, pp. 176–206.

Batley, Richard (2001), 'Public-private partnerships for urban services', in Freire, M., and R. Stren (eds.), *The Challenge of Urban Government: Policies and Practices*, Washington, DC: The World Bank, pp. 199–214.

Baylies, Carolyn, and Morris Szeftel (1997), 'The 1996 Zambian elections: still awaiting democratic consolidation', *Review of African Political Economy*, 24 (71), 113–128.

Beall, Jo, Owen Crankshaw, and Susan Parnell (2002), *Uniting a Divided City: Governance and Social Exclusion in Johannesburg*, London: Earthscan.

Becker, Charles, Andrew Hamer, and Andrew Morrison (1994), *Beyond Urban Bias in Africa: Urbanization in an Era of Structural Adjustment*, Portsmouth, New Hampshire: Heinemann.

Bennis, Phyllis (2000), *Calling the Shots: How Washington Dominates Today's UN*, Brooklyn, New York: Interlink.

Bilgin, P., and Morton, A.D. (2002), 'Historicising representations of "failed states": beyond the cold-war annexation of the social sciences?', *Third World Quarterly*, **23** (1), 55–80.

Blaikie, Piers, and Harold Brookfield (1987), *Land Degradation and Society*, London: Methuen.

Blunt, Alison, and Cheryl McEwan (2002), *Postcolonial Geographies*, London: Continuum.

Brennan, James (2002), 'Nation, race and urbanization in Dar es Salaam, Tanzania, 1916–1976', unpublished Ph.D. dissertation, Northwestern University.

Briggs, John, and Ian Yeboah (2001), 'Structural adjustment and the contemporary sub-Saharan African city', *Area*, **33** (1), 18–26.

Brunn, Stanley, Jack Williams, and Donald Ziegler (2003), *Cities of the World: World Regional Urban Development*, 3rd edition, New York: Harper Collins.

Bryant, Raymond (1991), 'Putting politics first: the political ecology of sustainable Development', *Global Ecology and Biogeography Letters*, **1**, 164.

Bryant, Raymond, and Sinead Bailey (1997), *Third World Political Ecology*, London: Routledge.

Buchanon, Keith, and J. Pugh (1955), *Land and People in Nigeria: The Human Geography of Nigeria and Its Environmental Background*, London: University of London Press.

Buckingham-Hatfield, Susan, and Susan Percy (1999), *Constructing Local Environmental Agendas: People, Places, and Participation*, London: Routledge.

Bull, Mutumba (2003), 'The gender dimension of the 2001 Zambian elections', *African Social Research*, **45/46**, 85–123.

Burgess, Rod, Maris Carmona, and Theo Kolstee (1997), *The Challenge of Sustainable Cities: Neoliberalism and Urban Strategies in Developing Countries*, Atlantic Highlands, New Jersey: Zed Books.

Burra, Marco (2004), 'Land use planning and governance in Dar es Salaam: a case study from Tanzania', in Hansen, Karen and Vaa, Mariken (eds.), *Reconsidering Informality: Perspectives from Urban Africa*, Uppsala, Sweden: Nordic Africa Institute, pp. 143–157.

Burton, Andrew (2002), 'Adjutants, agents, intermediaries: the Native Administration in Dar es Salaam township, 1919–1961', in Burton, Andrew (ed.), *The Urban Experience in Eastern Africa c. 1750–2000*, Nairobi: British Institute in Eastern Africa, pp. 98–118.

Byekwaso, A.W.S. (1994), 'The role of community based organizations in land servicing: a case study of Tabata, Buguruni and Vingunguti', unpublished diploma project, University of Dar es Salaam, Dar es Salaam, Tanzania.

Cammack, Paul (2002), 'Neoliberalism, the World Bank, and the new politics of development', in Kothari, Uma and Martin Minogue (eds), *Development*

Theory and Practice: Critical Perspectives, Basingstoke: Palgrave, pp. 157–178.

Central Statistical Office (CSO) (1998), *Living Conditions in Zambia–1998*, Lusaka: CSO.

CSO (2002), 2000 Census of Population and Housing: Presentation of Selected Indicators, Lusaka: CSO.

CSO (2003), *Lusaka Population by Residential Distribution*, prepared for author by CSO, Lusaka.

Chachage, Chachage S. (2000), *Environment, Aid and Politics in Zanzibar*, Dar es Salaam, Tanzania: University of Dar es Salaam Press.

Cheatle, Marion (1986), 'Water supply, sewerage, and drainage', in Williams, G. (ed.), *Lusaka and its Environs*, Lusaka: Zambia Geographical Association, pp. 250–8.

Cheelo, Kingsley Haanyembe (2002), 'Effects of community participation in urban water provision in Ng'ombe Township, Lusaka', unpublished BA research project report, University of Zambia.

Chege, Michael (1994), 'The return of multiparty politics', in Barkan, Joel (ed.), *Beyond Capitalism vs. Socialism in Kenya and Tanzania*, Boulder, Colorado: Lynne Rienner, pp. 47–74.

Cheru, Fantu (2002), *African Renaissance: Roadmaps to the Challenge of Globalization*, London and New York: Zed Books.

Chidumayo, Emmanuel (2002), 'Environmental issues in Zambia', *African Social Research*, **43/44**, 32–43.

Chileshe, Jonathan (1998) *Alderman Safeli Hannock Chileshe: A Tribute to the Man, His Life and History*, Ndola: Mission Press.

Chinamo, Elias (2003), Author interview with Elias Chinamo, Solid Waste Coordinator, Dar es Salaam City Council, June 10, 2003.

Chintowa, Paul (1995), 'Opposition queries CCM presidential victory', *InterPress Third World New Service*, 26 October 1995.

Chipimo-Mbizule, C. and B. Nundwe (1997), *Lusaka Longitudinal Livelihood Cohort Study: Results of a Baseline Study in Peri-Urban Lusaka*, Lusaka: Institute for Social and Economic Research.

Chonya, Morgan (2004), 'Extra budgetary spending affected poverty reduction, says IMF', *Zambian Mail*, July 12, 2004, p. 1.

Christie, James (1876), *Cholera Epidemics in East Africa*, London: Macmillan.

Cisse, Oumar (1996), 'Participatory solid waste management: urban community of Dakar, Senegal', in Gilbert, R., D. Stevenson, H. Girardet, and R. Stren (eds), *Making Cities Work*, London: Earthscan.

Collins, J. (1977), 'Lusaka: urban planning in a British colony', in Cherry, G. (ed.) *Shaping an Urban World*, New York: St. Martins Press, pp. 227–241.

Collins, J. (1986), 'Lusaka: the historical development of a planned capital, 1931–1970', in Williams, G. (ed.), *Lusaka and its Environs*, Lusaka: Zambia Geographical Association, pp. 95–137.

Cumming-Bruce, A-P (1954), Memo of Housing Officer A-P Cumming Bruce to Chief Secretary, Zanzibar National Archive file DA 1/261, Town Planning Legislation, 1954–1964.

Dahiya, Bharat, and Cedric Pugh (2000), 'The localization of Agenda 21 and the Sustainable Cities Programme', in Pugh, Cedric (ed.), *Sustainable Cities in Developing Countries*, London: Earthscan, pp. 152–184.

Danida (2002a), *The Power of Culture: The Cultural Dimension in Development*, Copenhagen: Royal Danish Ministry of Foreign Affairs.

Danida (2002b), *Country Strategy for Tanzania 2001–2005*, Copenhagen: Royal Danish Ministry of Foreign Affairs.

Danida (2003), *A World of Difference: The Government's Vision for New Priorities in Danish Development Assistance 2004–2008*, Copenhagen: Royal Danish Ministry of Foreign Affairs.

Danish Ministry of Foreign Affairs (2003), *Danish Development Policy*, Copenhagen: Royal Danish Ministry of Foreign Affairs.

Dar es Salaam City Council (1994), 'Subsidiary Legislation', Dar es Salaam: Government Printers.

Dar es Salaam City Commission (1999), *Strategic Urban Development Planning Framework*, Dar es Salaam: Dar es Salaam City Commission.

Darkoh, Michael (1994), *Tanzania's Growth Centre Policy and Industrial Development*, Berlin: Peter Lang.

Dear, Michael (2001), *From Chicago to L.A.: Making Sense of Urban Theory*, Thousand Oaks, California: Sage.

De Blij, Harm (1963), *Dar es Salaam: a Study in Urban Geography*, Evanston: Northwestern University Press.

Desai, Nitin (1999), 'Cultivating an urban eco-society: the United Nations response', in Inoguchi, Takashi, Edward Newman, and Glen Paoletto (eds), *Cities and the Environment*, Tokyo: United Nations University Press, pp. 243–255.

Devas, Nick, and Carole Rakodi (eds) (1993), *Managing Fast Growing Cities*, New York: John Wiley.

Diakhite, Haroua, and Jean-Louis Margerie (1996), 'Municipal policy and participatory dynamics: Kayes, Mali', in Gilbert, R., D. Stevenson, H. Girardet, and R. Stren (eds), *Making Cities Work*, London: Earthscan.

Djokotoe, Edem (2004), 'Prostitutes' play hits stage', *Zambian Post*, July 11, 2004, p. 15.

Doe, Ben, and Doris Tetteh (1999), 'The working group approach to environmental management under the Accra Sustainable Programme', Atkinson, Adrian, Julio Davila, Edesio Fernandes, and Michael Mattingly (eds), *The Challenge of Environmental Management in Urban Areas*, Brookfield, Vermont: Ashgate, pp. 171–179.

Dorman, W. Judson (2002), 'Authoritarianism and sustainability in Cairo: what failed urban development projects tell us about Egyptian politics', Zetter, Roger, and Rodney White (eds), *Planning in Cities: Sustainability and Growth in the Developing World*, ITDG Publishing, pp. 146–166.

Drakakis-Smith, David (1995), 'Third World cities: sustainable urban development I', *Urban Studies* **32** (4–5), 659–677.

Drakakis-Smith, David (1997), 'Third world cities: sustainable urban development III: basic needs and human rights', *Urban Studies* **34**, 797–823.

Drakakis-Smith, David (2000), *Third World Cities*, 2nd edition, London: Routledge.

Driver, Felix (2001), *Geography Militant: Cultures of Exploration and Empire*, Oxford: Blackwell.

Dutton, Eric (1948) Memo of Chief Secretary Eric Dutton to Director of Public Works, in Zanzibar National Archives file AB 76/72, Development and the PWD.

Eigen, Jochen (1998), 'Sustainable cities and local governance: lessons from a global UN program', in Fernandes, Edesio (ed.), *Environmental Strategies for Sustainable Development in Urban Areas: Lessons from Africa and Latin America*, Brookfield, Vermont: Aldershot, pp. 155–162.

Elmahi, Elmalieh M. (1993), 'The impact of Vingunguti and Tabata waste disposal sites on water quality of the area', unpublished MA thesis, University of Dar es Salaam, Dar es Salaam, Tanzania.

Environmental Council of Zambia (ECZ) (2001), *State of Environment in Zambia 2000*, Lusaka: ECZ.

ECZ and Lusaka City Council (1997), *Solid Waste Management Master Plan Project for the City of Lusaka*, Lusaka: ECZ and LCC.

Eriksen, Tore (2000), *Norway and National Liberation in Southern Africa*, Uppsala, Sweden: Nordic Africa Institute.

Escobar, Arturo (1995), *Encountering Development: the Making and Unmaking of the Third World*, Princeton: Princeton University Press.

Escobar, Arturo (2001), 'Culture sits in places: reflections on globalism and subaltern strategies of localization', *Political Geography*, **20**, 139–174.

Fair, Laura (2001), *Pastimes & Politics: Culture, Community, and Identity in Post-Abolition Urban Zanzibar, 1890–1945*, Athens, Ohio: Ohio University Press.

Fakih, Said (1995), Author interview with Acting Lands Officer Said Omar Fakih, June 22, 1995.

Faya, G. (2003), '30% of Dar residents have no toilets', *Guardian*, 29 May 2003, p. 2.

Fereji, Suleiman (1992), Author's interview with zanzibar resident Suleiman Farakhan Fereji, March 1, 1992.

Ferguson, James (1999), *Expectations of Modernity: Myths and Meanings of Urban Life on the Zambian Copperbelt*, Berkeley: University of California Press.

Fischer, F., and J. Forester (1999), *The Argumentative Turn in Policy Analysis and Planning*, London: University College London Press.

Foucault, Michel (1997), *The Essential Works 1954–1984, Vol. 1: Ethics, Subjectivity and Truth*, Rabinow, Paul (ed.), New York: The New Press.

Freidberg, Susanne (2001), 'Gardening on the edge: the social conditions of unsustainability on an African urban periphery', *Annals of the Association of American Geographers*, **91** (2), 349–369.

Freund, Bill (2001), 'Brown and green in Durban: the evolution of environmental policy in a post-apartheid city', *International Journal of Urban and Regional Research*, **25** (4), 717–739.

Friedmann, John (1998), 'The new political economy of planning: the rise of civil society', in Douglass, Mike, and John Friedmann (eds), *Cities for Citizens*, New York: Wiley, pp. 19–38.

Gandy, Matthew (2002), *Concrete and Clay: Reworking Nature in New York City*, Cambridge, Massachusetts: MIT Press.

German Development Service (DED) (2004), website of the German Development Service in Tanzania, at http://www.ded-tanzania.de.

German Embassy, Dar es Salaam (2004), website of the German Embassy in Tanzania, at http://www.german-embassy-daressalaam.de/

Gesellschaft fur Technische Zusammenarbeit (2004), website of the German Society for Technical Cooperation (GTZ), at http://www.gtz.de.

Ghassany, Mohammed (2003a), 'Saa ya ukombozi inarejea!' ['The hour of liberation is returning!'], *Dira*, 20–26 June 2003, p. 4.

Ghassany, Mohammed (2003b), 'Hakuna Ugozi, hakuna Uhizbu, kuna Uzanzibari' ['There's no Gozi-ness, there's no Hizbu-ness, there is Zanzibari-ness'], *Dira*, 27 June–3 July 2003, p. 4.

Ghassany, Mohammed (2003c), 'Maalim Shamsi: hoja sio tena Zanzibar kumezwa, ni kumezuliwa!' ['Professor Shamsi: the concern is no longer about Zanzibar being swallowed, but being unswallowed'], *Dira*, 4–10 July 2003, p. 4.

Ghassany, Mohammed (2003d), 'Ya Raza na ya Feruzi: nusu uongo, nusu ukweli' ['(Claims) of Raza and of Feruzi: half right, half wrong'], *Dira*, 25–31 July 2003, p. 4.

Ghassany, Mohammed (2003e), 'Uzanzibari si ngoma tu' ['Zanzibari-ness isn't just about dances'], *Dira*, 1–7 August 2003, p. 4.

Ghassany, Mohammed (2003f), 'Watawala ni raia: hekaya ya Lila na Fila' ['The rulers are the citizens: the legend of Lila and Fila'], *Dira*, 8–14 August 2003, p. 4.

Girardet, H. (1999), *Creating Sustainable Cities*, Devon: Green Books.

Gossling, Stefan (2002), 'The political ecology of tourism in Zanzibar', in Gossling, Stefan (ed.), *Tourism and Development in Tropical Islands: Political Ecology Perspectives*, London: Edward Elgar.

Gregory, Derek (2004), *The Colonial Present: Afghanistan, Palestine, and Iraq*, Malden, Massachusetts: Blackwell.

Griffin, Angela (2001), 'The promotion of sustainable cities', in Freire, M. and R. Stren (eds), *The Challenge of Urban Government: Policies and Practices*, Washington, DC: The World Bank, pp. 63–72.

Guardian (2003), 'Tanzanian minister gagged, robbed of 80 million shillings', *Guardian*, 16 July 2003, p. 1.

Habeenzu, Mulunda (2002), 'Kabungo defends demolitions', *Zambian Post*, December 13, 2002, p. 7.

Halfani, Mohamed (1996), 'Marginality and dynamism: prospects for the Sub-Saharan African city', in Cohen, M., B. Ruble, J. Tulchin, and A. Garland (eds), *Preparing for the Urban Future*, Washington, DC: Woodrow Wilson Center Press, pp. 83–107.

Halfani, Mohamed (1997a), 'Governance of urban development in East Africa: an examination of the institutional landscape and the poverty challenge', in

Swilling, Mark (ed.), *Governing Africa's Cities*, Johannesburg: Witwatersrand University Press, pp. 115–159.

Halfani, Mohamed (1997b), 'The challenge of urban governance in Africa: institutional change and the knowledge gaps', in Swilling, Mark (ed.), *Governing Africa's Cities*, Johannesburg: Witwatersrand University Press, pp. 13–34.

Halla, Francos (1994), 'A coordinating and participatory approach to managing cities: the case of the Sustainable Dar es Salaam Project in Tanzania', *Habitat International*, **18**, 19–31.

Halla, Francos (1997), 'Institutional arrangements for urban management: the Sustainable Dar es Salaam Project', unpublished doctoral dissertation, Rutgers University, New Brunswick, New Jersey.

Halla, Francos (1998), 'Engaging in environmental planning and management: a false start in Dar es Salaam', *Journal of Building and Land Development*, **5** (3), 62–71.

Halla, Francos, and Bituro Majani (1999a), 'The EPM process and the conflict over outputs in Dar es Salaam', *Habitat International*, **23** (3), 339–350.

Halla, Francos, and Bituro Majani (1999b), 'Innovative ways for solid waste management in Dar es Salaam: toward stakeholder partnerships', *Habitat International*, **23** (3), 351–361.

Halla, Francos (2003), Personal communication with author, June 5, 2003.

Ham, M. (1992), 'Luring investment', *Africa Report*, September/October, 39–41.

Hamadi, Ali (2003), Comments of Ali Hamadi of the Water Supply Working Group, Zanzibar Sustainable Program, in a focus group discussion with author, July 9, 2003.

Hance, William (1964), *The Geography of Modern Africa*, New York: Columbia University Press.

Hansen, Karen (1997), *Keeping House in Lusaka*, New York: Columbia University Press.

Hansen, Karen (2000), *Salaula: The World of Secondhand Clothing and Zambia*, Chicago: University of Chicago Press.

Hansen, Karen, and Mariken Vaa (2004), *Reconsidering Informality: Perspectives from Urban Africa*, Uppsala, Sweden: Nordic Africa Institute.

Hanson, Susan, and Robert Lake (2000), 'Needed: geographic research on urban sustainability', *Urban Geography*, **21** (1), 1–4.

Hardoy, Jorge, D. Mitlin, and D. Satterthwaite (1992), *Environmental Problems in Third World Cities*, London: Earthscan.

Hayward, Bronwyn (2003), 'Deliberative planning and urban sustainability', in Freeman, Claire, and Michelle Thompson-Fawcett (eds), *Living Space: Towards Sustainable Settlements in New Zealand*, Dunedin, New Zealand: University of Otago Press, pp. 113–130.

Henry, Nick, Jane Pollard, and James Sidaway (2001), beyond the margins of economics: geographers, economists, and policy relevance. *Antipode*, **41** (3), 200–207.

Herbst, Jeffrey (2000), *States and Power in Africa: Comparative Lessons in Authority and Control*, Princeton: Princeton University Press.

Holm, Mogens (1995), 'The impact of structural adjustment on intermediate towns and urban migrants: an example from Tanzania', in Simon, D., W. van Spengen, C. Dixon, and A. Narman (eds), *Structurally Adjusted Africa*, London: Pluto Press, pp. 91–106.

Holston, James (1995), 'Spaces of insurgent citizenship', in Sandercock, Leonie (ed.), *Making the Invisible Visible: New Historiographies for Planning*, Milan, Italy: Franco Angeli, pp. 35–52.

Hoogvelt, Ankie (1997), *Globalization and the Postcolonial World: The New Political Economy of Development*, Baltimore: Johns Hopkins University Press.

Huntington, Samuel, and E. Harrison (2000), *Culture Matters: How Values Shape Human Progress*, Cambridge: Harvard University Press.

Hussain, Tabasim (1992), 'End of Tanzania's one party rule', *Africa Report*, July/August 1992, pp. 22–23.

Hussein, Asha (2003), 'Abode of vagabonds in making at Ilala Municipality', *Guardian*, 3 June 2003, p. 6.

Hyden, Goran (1994), 'Party, state, and civil society: control versus openness', in Barkan, Joel (ed.), *Beyond Capitalism vs. Socialism in Kenya and Tanzania*, Boulder, Colorado: Lynne Rienner, pp. 75–99.

Hyden, Goran, and John Mugabe (1999), 'Governance and sustainable development in Africa: the search for economic and political renewal', in Mugabe, John (ed.), *Governing the Environment: Political Change and Natural Resources Management in Eastern and Southern Africa*, Nairobi: ACTS Press, pp. 29–38.

Ihonvbere, Julius (1996), *Economic Crisis, Civil Society, and Democratization: the Case of Zambia*, Trenton: Africa World Press.

Inoguchi, Takashi, Edward Newman, and Glen Paoletto (1999), 'Introduction: cities and the environment – towards ecopartnerships', in Inoguchi, Takashi, Edward Newman, and Glen Paoletto (eds), *Cities and the Environment*, Tokyo: United Nations University Press, pp. 1–14.

Insight [Quarterly Newspaper of the Sustainable Lusaka Program] (1999a), 'LCC needs financial surgery', *Insight* vol. 5, March 31, 1999, p. 1.

Insight (1999b), 'New mayor embarks on house demolition errand', *Insight* vol. 5, March 31, 1999, p. 1.

Insight (1999c), 'Update on communities', *Insight* vol. 5, March 31, 1999, p. 1.

International Labor Organization (ILO) (1998), 'Evaluation of Hanna Nassif community based urban upgrading programme in Dar es Salaam', Dar es Salaam: ILO.

Ireland Aid Review Committee (2002), *Report of the Ireland Aid Review Committee*, Dublin: Ireland Aid.

Ivaska, Andrew (2003), 'Negotiating culture in a cosmopolitan capital: urban style and the Tanzanian state in colonial and postcolonial Dar es Salaam', unpublished Ph.D. Dissertation, University of Michigan.

Jenkins, Paul, and Harry Smith (2002), 'International agency shelter policy in the 1990s: experience from Mozambique and Costa Rica', Zetter, Roger, and Rodney White (eds), *Planning in Cities: Sustainability and Growth in the Developing World*, ITDG Publishing, pp. 131–145.

Jobo, Mohamed (1992), Author's interview with Zanzibar resident Mohamed Ali Jobo, July 11 1992.

Juma, Sheha Mjaja (2003), Author interview with Sheha Mjaja Juma, project director, Zanzibar Sustainable Program, June 26 2003.

Kabuba, Ireen (2002), Interview by author and Wilma Nchito with Ireen Kabuba, former director, Sustainable Lusaka Program, December 15, 2002.

Kaindu, Maluba (2002), 'Lusaka Council shortlists five firms to collect garbage', *Times of Zambia*, December 12 2002, p. 1.

Kaiser, Paul (1996), 'Structural adjustment and the fragile nation: the demise of social unity in Tanzania', *Journal of Modern African Studies*, **34** (2), 227–37.

Kalapula, E. S. (1994), 'Urbanization, urban development planning and management in Zambia: an overview', in Wekwete, K. and C. Rambanapasi (eds.), *Planning Urban Economies in Southern and Eastern Africa*, Aldershot: Avebury, pp. 55–75.

Kamanga, Zakeyo (1989), 'The causes and effects of soil erosion in shanty compounds: a case study of Ng'ombe compound', unpublished Bachelor of Arts research project report, University of Zambia.

Kapwepwe, Mulenga (2004), *Like Choosing Between Eating and Breathing*, play performed at the Lusaka Playhouse, July 15, 2004.

Karp, Ivan (1997) 'Does theory travel? Area studies and cultural studies'. *Africa Today*, **44**, 281–296.

Kashoki, Mubanga, and Mwale, Stephen (2003), 'The 2001 Zambian presidential, parliamentary, and local government elections: education of the electorate', *African Social Research*, **45/46**, 46–84.

Kasumuni, L. (2003), 'Half of all Dar residents use filthy water', *Guardian*, 11 June 2003, p. 5.

Kaswende, Kingsley (2004), 'Cost of living highest in Luanshya – JCTR', *Zambian Post*, July 14 2004, p. 9.

Keis, Ahmed (2003), Opening address by Zanzibar Mayor Ahmed Keis, mayoral forum with city leaders on solid waste issues, August 14, 2003.

Kelsall, Tim (2004), *Contentious Politics, Local Governance and the Self: a Tanzanian Case Study*, Uppsala, Sweden: Nordic Africa Institute.

Kharusi, Ahmed (1967), *Zanzibar, Africa's First Cuba: A Case Study of the New Colonialism*, Richmond, UK: Foreign Affairs Publishing.

Kiondo, Andrew (1994), 'The new politics of local development in Tanzania', in Kanyinga, K., A. Kiondo and P. Tidemand (eds), *The New Local Level Politics in East Africa*, Uppsala, Sweden: Nordic Africa Institute, pp. 50–88.

Kiondo, Andrew (1995), 'When the state withdraws: local development, politics and liberalization in Tanzania', in Gibbon, Peter (ed.), *Liberalized Development in Tanzania*, Uppsala, Sweden: Nordic Africa Institute, pp. 109–176.

Kipfer, Stefan (1996), 'Whose sustainability? Ecology, hegemonic politics and the future of the city', in Keil, Roger, Gerda Wekerle, and David Bell (eds), *Local Places in the Age of the Global City*, Montreal: Black Rose Books, pp. 117–124.

Kironde, J.M.L (1999), 'The governance of waste management in African cities', in Atkinson, Adrian, Julio Davila, Edesio Fernandes, and Michael Mattingly (eds), *The Challenge of Environmental Management in Urban Areas*, Brookfield, Vermont: Ashgate, pp. 75–88.

Kironde, J.M.L. (2001a), *Financing the Sustainable Development of Cities in Tanzania: the Case of Dar es Salaam and Mwanza*, Dar es Salaam: University College of Lands and Architectural Studies.

Kironde, J.M.L. (2001b), *Urban Poverty in Tanzania: the Role of Urban Authorities*, Dar es Salaam: University College of Lands and Architectural Studies.

Kironde, J.M.L. and S. Ngware (2000), 'Introduction', in Ngware, S., and J.M.L. Kironde (eds), *Urbanising Tanzania: Issues, Initiatives, and Priorities*, Dar es Salaam: University of Dar es Salaam Press. pp. 1–6.

Kishimba, M.A. and A. Mkenda (1995), 'The impact of structural adjustment programs on urban pollution and sanitation: empirical evidence from urban centers in Tanzania', in Bagachwa, M.S., and Festus Limbu (eds), *Policy Reform and the Environment in Tanzania*, Dar es Salaam: University of Dar es Salaam Press, pp. 197–225.

Kitilla, Martin (1999), 'Strategic urban development plan for Dar es Salaam', in Caldwell, Wayne (ed.), *Issues and Responses: Land Use Planning in Eastern and Southern Africa*, Harare, Zimbabwe: Weaver Press, pp. 117–134.

Kitilla, Martin (2001), 'Tanzania national programme – national Sustainable Cities Programme: Urban Authorities Support Unit (UASU): replicating EPM country-wide', in *Implementation and Replication of the Sustainable Cities Programme Process at City and National Level: Case Studies from Nine Cities*, Nairobi, Kenya: United Nations Environment Programme and United Nations Habitat, pp. 86–106.

Kitilla, Martin (2003), Author interview with Martin Kitilla, Director, Urban Authorities Support Unit in the Dar es Salaam City Council, June 19, 2003.

Knauder, Stefanie (1982), *Shacks and Mansions: An Analysis of the Integrated Housing Policy in Zambia*, Lusaka, Zambia: Multimedia Publishers.

Komakoma, Joe (2003), 'The road to nowhere', *Zambian Post*, January 9, 2003, p. 8.

Kombe, Wilbard (1994), 'The demise of public urban land management and the emergence of informal land markets in Tanzania: a case of Dar-es-Salaam City', *Habitat International*, **18** (1), 23–43.

Kombe, Wilbard (2001), 'Institutionalising the concept of environmental planning and management (EPM): successes and challenges in Dar es Salaam', *Development in Practice*, **11** (2/3), 190–207.

Kondoro, J.W. (1995), 'The impact of SAPs on urban and industrial pollution in Tanzania: the case study of Dar es Salaam city', in Bagachwa, M.S., and Festus Limbu (eds), *Policy Reform and the Environment in Tanzania*, Dar es Salaam: University of Dar es Salaam Press, pp. 227–249.

Kothari, Uma, and Martin Minogue (2002), 'Critical perspectives on development: an introduction', in Kothari, Uma, and Martin Minogue (eds), *Development Theory and Practice: Critical Perspectives*, Basingstoke: Palgrave, pp. 1–15.

Kreditanstalt fur Wiederaufbau (KFW) [German Development Bank] (1998), *Solid Waste Management of Zanzibar Town: Rehabilitation and Improvement of Zanzibar Municipality's Sewerage, Drainage, and Solid Waste Disposal System: Final Report*, compiled by Peter Braun, Hamburg, Germany: KFW.

Kulaba, Saitiel (1989), 'Local government and the management of urban services in Tanzania', in Stren, R., and R. White (eds), *African Cities in Crisis*, Boulder: Westview, pp. 203–245.

Kyessi, A. (1998), 'City expansion and urban agriculture in Dar es Salaam: lessons for planning', *Journal of Building and Land Development*, **5** (2), 9–17.

Kyessi, A. (2001), 'Community-based urban water management under scarcity in Dar es Salaam, Tanzania', *Journal of Building and Land Development*, **8** (1–3), 28–41.

Kyessi, A.G. and S.A. Sheuya (1993), 'The role of the community based organizations and non-governmental organizations in squatter upgrading', paper presented to the 5th International Seminar on Construction Management for Sustainable Self-Help Housing in Habinet Countries, Dar es Salaam, 12–21 October 1993.

Lerise, Fred (2000), 'Urban governance and urban planning in Tanzania', in Ngware, S., and J.M.L. Kironde (eds), *Urbanising Tanzania: Issues, Initiatives, and Priorities*, Dar es Salaam: University of Dar es Salaam Press, pp. 88–116.

Lerise, Fred (2003), Personal communication with Fred Lerise, August 12 2003.

Lerise, Fred, and S. Ngware (2000), 'Managing urban development in Tanzania: issues, initiatives and priorities', in Ngware, S., and J.M.L. Kironde (eds), *Urbanising Tanzania: Issues, Initiatives, and Priorities*, Dar es Salaam: University of Dar es Salaam Press, pp. 117–132.

Lewinson, Anne (1999), 'Going with the times: transforming visions of urbanism and modernity among professionals in Dar es Salaam, Tanzania', unpublished Ph.D. dissertation, University of Wisconsin, Madison.

Light, Richard (1941), *Focus on Africa*, New York: American Geographical Society.

Lofchie, Michael (1965), *Zanzibar: Background to Revolution*, Princeton: Princeton University Press.

Loomba, A. (1998), *Colonialism/Postcolonialism*, London: Routledge.

Lorenz, Nicolaus, and Deo Mtasiwa (2003), 'Health in the urban environment – experience from Dar es Salaam, Tanzania', paper presented to the New York Academy of Sciences Conference on Urban Biosphere and Society, 29–30 October 2003.

Lubelski, Marek, and Raff Carmen (1999), 'The North-South dimension of LA21', in Buckingham-Hatfield, Susan, and Susan Percy (eds), *Constructing Local Environmental Agendas: People, Places, and Participation*, London: Routledge, pp. 110–22.

Lugalla, Joe (1995), *Adjustment and Poverty in Tanzania*, Munster, Germany: Lit Verlag.

Lugalla, Joe, and Colleta Kibassa (2003), *Urban Life and Street Children's Health: Children's Accounts of Urban Hardships and Violence in Tanzania.* Hamburg: Lit Verlag.

Mabin, Alan (2001), 'Contested urban futures: Report on a global gathering in Johannesburg, 2000'. *International Journal of Urban and Regional Research*, **25**, (1), 180–184.

Mabogunje, Akin (1968), *Urbanization in Nigeria*, London: University of London Press.

Maimbo, Mutinta (2004), 'MMD, UPND clash at Lwiindi ceremony', *Zambian Post*, July 9, 2004, p. 11.

Main, Hamish and Steven Williams (1994), *Environment and Housing in Third World Cities*, New York: John Wiley.

Maira, Julius (2001), 'Sustainable Dar es Salaam Programme, Tanzania – evolvement, development, and experiences in implementing environmental strategies through investments and policy initiatives', in *Implementation and Replication of the Sustainable Cities Programme Process at City and National Level: Case Studies from Nine Cities*, Nairobi, Kenya: United Nations Environment Programme and United Nations Habitat, pp. 35–50.

Majani, Bituro (1998), 'Transaction costs and institutional change in solid waste management: a study of the scavengers of the Vingunguti refuse dump, Dar es Salaam', *Journal of Building and Land Development*, **5** (3), 1–14.

Majani, Bituro (2000), 'Institutionalizing environmental planning and management: the institutional economics of solid waste management in Tanzania', unpublished PhD dissertation, Dortmund, Germany: University of Dortmund.

Majani, Bituro (2002), 'Environmental planning and management in Dar es Salaam: a global success story and learning experience', *Journal of Building and Land Development*, **9** (1), 67–72.

Majani, Bituro (2003a), 'Improving the urban environment: from doing things right to doing the right things', *Journal of Building and Land Development*, **10** (1), 1–16.

Majani, Bituro (2003b), Author interview with Bituro Majani, former director, Sustainable Dar es Salaam Project, June 5, 2003.

Majira (2003), 'Editorial: hili la makaburi liangaliwe upya' ['This matter of the graves must be looked at anew'], *Majira*, 10 August 2003, p. 6.

Malamba, Joseph (2004), Personal communication with Pastor Joseph Malanga of Garden compound, July 13, 2004.

Malera, M. (2003), 'Waislamu wateka makaburi' ['Muslims storm the graves'], *Majira*, 9 August 2003, p. 1.

Maliyamkono, T. (2000), 'Contemporary Zanzibar: a survey of facts and views', in Maliyamkono, T. (ed.), *The Political Plight of Zanzibar*, Dar es Salaam: Tema Publishers, pp. 153–175.

Mamdani, Mahmood (1996), *Citizen and Subject: Contemporary Africa and the Legacy of Late Colonialism*, Princeton, New Jersey: Princeton University Press.

Mandrad, Anthony (1992), Author's interview with Zanzibar resident Anthony Thomas Mandrad, May 16, 1992.

Mapuri, Omar (1996), *Zanzibar, The 1964 Revolution: Achievements and Prospects*, Dar es Salaam: Tema Publishers.

Martin, Richard (1975), 'The evolution of a traditional morphology in an urban setting: Greater Lusaka', *Zambian Geographical Journal*, **29–30**, 1–19.

Massaro, Richard (1998), 'The political economy of spatial rationalization and integration policies in Tanzania', in Silberfein, M. (ed.), *Rural Settlement Structure and African Development*, Boulder, Colorado: Westview Press, pp. 273–307.

Massey, Doreen (2001), 'Geography on the agenda', *Progress in Human Geography*, **25** (1): 5–17.

Matavika, G. (2003), 'Effects of globalisation on Tanzanian culture', *Guardian*, 2 June 2003, p. 11.

Mate, Litumelo (2001), *Sustainable Lusaka Programme ZAM/97/002: Programme Report*, Lusaka: Ministry of Local Government and Housing, Republic of Zambia.

Mate, Litumelo (2002), Interview by author and Wilma Nchito with Litumelo Mate, former Director, Sustainable Lusaka Program, December 16, 2002.

Mato, Rubhera (2002), 'Groundwater pollution in urban Dar es Salaam, Tanzania: assessing vulnerability and protection priorities', unpublished PhD dissertation, Eindhoven Technical University, Eindhoven, The Netherlands.

Mbembe, Achille (2001), *On the Postcolony*, Berkeley: University of California Press.

Mbewe, Mzamose (2002), Interview with Samalila Ukhondo Waste Group secretary Mzamose Mbewe, conducted by author and Wilma Nchito, December 23, 2002.

Mbogoni, Lawrence (2004), *The Cross versus The Crescent: Religion and Politics in Tanzania from the 1880s to the 1990s*, Dar es Salaam: Mkuki na Nyota Publishers.

Mbwiliza, J. (2000), 'The birth of a political dilemma and the challenges of the quest for new politics in Zanzibar', in Maliyamkono, T. (ed.), *The Political Plight of Zanzibar*, Dar es Salaam: Tema Publishers, pp. 1–34.

McCann, Eugene (2001), 'Collaborative visioning or urban planning as therapy? The politics of public-private policy-making', *The Professional Geographer*, **53** (2), 207–218.

Meebelo, Henry (1973), *Main Currents of Zambian Humanist Thought*, Lusaka: Oxford University Press.

McEwan, Cheryl (2003a), 'The West and other feminisms', in Anderson, K., M. Domosh, S. Pile, and N. Thrift (eds), *Handbook of Cultural Geography*, London: Sage, pp. 405–19.

McEwan, Cheryl (2003b), 'Material geographies and postcolonialism', *Singapore Journal of Tropical Geography*, **24** (3), 340–55.

McGranahan, Gordon, and David Satterthwaite (2002), 'The environmental dimensions of sustainable development for cities', *Geography*, **87** (3), 213–226.

Menda, Aloyce (1991), 'Hamad tortured, Isles elders claim', *Business Times*, 4 October 1991, p. 1.

Mercer, Claire (1999), 'Reconceptualizing state-society relations in Tanzania: are NGOs "Making a Difference"?', *Area*, **31** (3), 247–258.

Mercer, Claire (2003), 'Performing partnership: civil society and the illusions of good governance in Tanzania', *Political Geography*, **22**, 741–63.

Mhamba, Robert, and Colman Titus (2001), 'Reactions to deteriorating provision of public services in Dar es Salaam', in Tostensen, Arne, Inge Tvedten, Mariken Vaa (eds), *Associational Life in African Cities: Popular Responses to the Urban Crisis*, Uppsala, Sweden: Nordic Africa Institute, pp. 218–231.

Middleton, Neil, and Phil O'Keefe (2003), *Rio Plus Ten: Politics, Poverty, and the Environment*, London: Pluto.

Minogue, Martin (2002), 'Power to the people? Good governance and the reshaping of the state', in Kothari, Uma and Martin Minogue (eds), *Development Theory and Practice: Critical Perspectives*, Basingstoke: Palgrave, pp. 117–135.

Mitchell, Don (2000), *Cultural Geography: a Critical Introduction*, Malden, Massachusetts: Blackwell.

Mkandawire, Peter (2002), Personal communication with Ng'ombe resident Peter Mkandawire, December 19 2002.

Mkumba, Michael (2003), Author interview with Michael Mkumba, landfill manager for Dar es Salaam City Council, June 11 2003.

Mlambo, Asteria (2003), Author interview with Asteria Mlambo, EPM Coordinator, Dar es Salaam City Council, June 9 2003.

Mloo, Shaaban (1964), Memo of Liaison Officer Shaaban Mloo to Chief Minister, Zanzibar National Archives file DA 2/5, DDR Housing Scheme, 1964–1968.

Mmuya, Maximilian (1998), *Tanzania: Political Reform in Eclipse: Crises and Cleavages in Political Parties*, Dar es Salaam, Tanzania: Friedrich Ebert Stiftung.

Mnoga, Abdulrahman (2003), Comments of former Zanzibar mayor Abdulrahman Mnoga in a focus group discussion with author, July 9, 2003.

Mohammed, Abdullah (2001), 'State-civil society relationship for democracy and sustainable development: a case study of municipal governance in Zanzibar Town', unpublished MA thesis, University of Dar es Salaam, Dar es Salaam, Tanzania.

Mohammed, Juma (2003), 'Manispaa Z'bar yaambiwa "kajifunzeni kwa Keenja"' ['Zanzibar Municipal Council is told "go and learn from Keenja"'], *Rai*, 14–20 August 2003, p. 19.

Mohan, G., and K. Stokke (2000), 'Participatory development and empowerment: the dangers of localism', *Third World Quarterly*, **21** (2), 247–268.

Moore, Donald (1993), 'Contesting terrain in Zimbabwe's Eastern Highlands: political ecology, ethnography, and peasant resource struggles', *Economic Geography*, **69**, 380–401.

Moore, Donald (1997), 'Remapping resistance: "ground for struggle" and the politics of Place', in Pile, Steve, and Michael Keith (eds), *Geographies of Resistance*, London: Routledge, pp. 87–106.

Mphuka, Chrispin (2002), 'An overview of Zambia's economy and poverty situation', *African Social Research*, **43/44**, 1–7.

Mpuya, Michael (2000), 'Urban poverty and squatting in marginal lands: a case study of Hananasif-Bondeni, Yombo-Kilakala, and Vingunguti-Msimbazi Valley in Dar es Salaam', unpublished MA thesis, University of Dar es Salaam, Dar es Salaam, Tanzania.

Mtani, Anna (2003), Personal communication with Anna Mtani, project coordinator, Safer Cities Program in Dar es Salaam, June 2, 2003.

Mudimbe, V.Y. (1994), *The Idea of Africa*, Bloomington, Indiana: Indiana University Press.

Muhajir, Makame (2003), Personal communication with Makame Ali Muhajir, former director of Urban Planning and Surveys, Zanzibar, August 11, 2003.

Muhajir, M.A., S.K. Khairi, and Sheha M. Juma (1998), 'Proposition paper on upgrading of Ng'ambo areas in the Zanzibar Municipality', unpublished proposition paper, Zanzibar Sustainable Program.

Muhsin, Ramadhan Juma (2003), Author interview with Rama J. Muhsin, Solid Waste Coordinator, Zanzibar Municipal Council, July 10, 2003.

Mukangara, D. (2000), 'Race, ethnicity, religion, and politics in Zanzibar', in Maliyamkono, T. (ed.), *The Political Plight of Zanzibar*, Dar es Salaam: Tema Publishers, pp. 35–54.

Mukuka, Lawrence (2001), 'A vision for the future of Zambia and Africa', in Olugbenga Adesida and Arunma Oteh (eds.), *African Voices, African Visions*, Uppsala, Sweden: Nordiska Afrikainstitutet, pp. 110–123.

Mulenga, Chileshe (2003), 'An alternative interpretation of recent multiparty election results in Africa and implications for consolidation of democracy: the case of Zambia', *African Social Research*, **47/48**, 1–20.

Mulenga, Chileshe, and Bwalya, Edgar (2003), 'Voter attitudes to the 2001 Zambian elections: before, during, and after', *African Social Research*, **45/46**, 124–150.

Mumba, Bernadette (2003), Author's interview with Bernadette Mumba, Environmental Health Officer, Lusaka Province, January 14 2003.

Mumbi, Steven (2002), Author's interview with Mtendere resident Steven Mumbi, December 30 2002.

Mumbi, Steven (2003), Author's interview with Mtendere resident Steven Mumbi, January 19 2003.

Mundia, Ruth (2003), Interview by author and Wilma Nchito with Ruth Mundia, Zaninge Waste Group, Ng'ombe, Lusaka, January 2 2003.

Mupuchi, Speedwell (2004a), 'Govt invites private sector to manage markets, bus stations', *Zambian Post*, July 11 2004, p. 2.

Mupuchi, Speedwell (2004b), 'Reduce taxes on production – ZAM', *Zambian Post*, July 12 2004, p. 11.

Mussa, Ussi (2003a), 'Kipindupindu charudi tena' ['Cholera returns again'], *Zanzibar Leo*, August 12 2003, p. 1.

Mussa, Ussi (2003b), 'Uhaba wa magari unakwamisha uzoaji taka' ['Truck shortage stalls garbage collection'], *Zanzibar Leo*, July 15 2003, p. 6.

Mutale, Emmanuel (2004), *The Management of Urban Development in Zambia*, Aldershot: Ashgate.

Mutale, Michael (2002), Interview by author and Wilma Nchito with Michael Mutale, GIS Officer, Lusaka City Council, December 19 2002.

Mutenga, C.R. and S.L. Muyakwa (1999), 'Livelihoods from solid waste in Lusaka City: challenges and opportunities', Lusaka: Environmental Support Program of the Ministry of Environment and Natural Resources.

Muwowo, Francis (2001), 'Sustainable Lusaka Project, Zambia – experiences in periurban environmental infrastructure demonstration projects', in *Implementation and Replication of the Sustainable Cities Programme Process at City and National Level: Case Studies from Nine Cities*, Nairobi: United Nations Environment Programme, pp. 30–34.

Mwanatongoni, Mbena (1991), 'Germans fully involved in Zanzibar', *Business Times* [Dar es Salaam], October 4 1991, p. 3.

Mwangu, Danny (2003), Author interview with Danny Mwangu, Hazardous Waste Inspector, Environmental Council of Zambia, January 23 2003.

Mwizabi, Gethsemane (2004), 'PUSH through 'food for work' strategy aims at poverty reduction', *Times of Zambia*, July 10 2004, p. 5.

Myers, Garth (1994a), 'Eurocentrism and African urbanization: the case of Zanzibar's Other Side', *Antipode*, **26** (3), 195–215.

Myers, Garth (1994b), 'Making the socialist city of Zanzibar', *Geographical Review*, **84** (4), 451–464.

Myers, Garth (1994c), 'From "Stinkibar" to "The Island Metropolis of Eastern Africa": the geography of British hegemony in Zanzibar', in Godlewska, Anne, and Neil Smith (eds), *Geography and Empire*, Malden, Massachusetts: Blackwell, pp. 212–227.

Myers, Garth (1996), 'Democracy and development in Zanzibar? Contradictions in land and environment planning', *Journal of Contemporary African Studies*, **14**, 221–245.

Myers, Garth (1998a), 'Intellectual of empire: Eric Dutton and hegemony in British Africa', *Annals of the Association of American Geographers*, **88** (1), 1–27.

Myers, Garth (1998b), 'Incorporating indigenous knowledge into African urban planning: a case study with Chinese connections', *Hong Kong Papers in Design and Development*, **1** (1), 46–55.

Myers, Garth (1999), 'Political ecology and urbanisation: Zanzibar's construction materials industry', *Journal of Modern African Studies*, **37** (1), 83–108.

Myers, Garth (2000), 'Narrative representations of revolutionary Zanzibar', *Journal of Historical Geography*, **26** (3), 429–448.

Myers, Garth (2002a), 'Local communities and the new environmental planning: a case study from Zanzibar', *Area*, **34** (2), 149–159.

Myers, Garth (2002b), 'Planning the sustainable city: towards a political ecology of urban growth in Zanzibar', in Sassen, S. (ed.), in *Human Resource System Challenge VII: Human Settlement Development: Encyclopedia of Life*

Support Systems (EOLSS), Oxford: UNESCO and EOLSS Publishers, available at http://www.eolss.net.

Myers, Garth (2003), *Verandahs of Power: Colonialism and Space in Urban Africa*, Syracuse: Syracuse University Press.

Myers, Garth, and Makame Muhajir (1997), 'Localising Agenda 21: environmental sustainability and Zanzibari urbanization', *Third World Planning Review*, **19** (4), 367–384.

Mzee, Fatma, and Juma Masoud (1992), 'Mfalme wa Sharja asema uhusiano wa Ghuba, Afrika uimarishwe' ['Sultan of Sharja says relations between Gulf, Africa, must be strengthened'], *Nuru*, 16 December 1992, p. 1.

Nabudere, Dani (2000), *Globalisation and the Post-Colonial African State*, Harare: SAPES Books.

Nagar, Richa (1997), 'The making of Hindu communal organizations, places, and identities in post-colonial Dar es Salaam', *Environment and Planning D: Society and Space*, **15**, 707–730.

Namoonde, Nduba (2004), 'High PAYE impacts on savings', *Zambian Post*, July 10, 2004, p. 7.

Narman, A., and D. Simon (1998), 'Introduction', in Simon, D. and A. Narman (eds), *Development as Theory and Practice: Current Perspectives on Development and Development Co-operation*, Harlow, Essex: Longman, pp. 1–13.

Nasaha (2000), 'Tume ya Keenja haikufanya ilichoagizwa – Wananchi' ['Keenja's Commission didn't do what it was ordered to do – citizens'], *Nasaha*, 5–11 April, 2000, p. 1.

National Archives of Zambia (NAZ), Index to Lusaka Urban District Council files 1/1–1/25: Administrative History.

NAZ, Local Government and Housing (LGH) file 1/13/26: Lusaka – Delimitation of Wards 1963.

NAZ, Lusaka Urban District Council (LUDC) file 1/4/24: African Suburbs and Hostels, Private Compounds and Townships: Unauthorized Compounds 1957–1961.

NAZ, LUDC file 1/10/10: Town Planning – Town Planning Authority (Peri-Urban) – Policy and Procedure 1954.

NAZ, LUDC file 1/4/40: African Suburbs – Private Compounds and Townships – Unauthorized Compounds 1961.

NAZ, LUDC file 1/4/47: African Suburbs Private Compounds and Townships: Unauthorized Compounds 1962.

NAZ, LUDC file 1/8/4: African Participation in Local Government 1955–1960, including Hansard 86h.

NAZ, LUDC file 1/18/4: Sanitation Services – Refuse Removal Service – Tipping Site, 1952–1965.

National Research Council of the National Academies (2002), *Down to Earth: Geographic Information for Sustainable Development in Africa*, Washington, DC: The National Academies Press.

Nchito, Wilma, and Myers, Garth (2004), 'Four caveats for participatory solid waste management in Lusaka', *Urban Forum*, **15** (2), 109–133.

Ndulu, Benno, and Francis Mwega (1994), 'Economic adjustment policies', in Barkan, Joel (ed.), *Beyond Capitalism vs. Socialism in Kenya and Tanzania*, Boulder, Colorado: Lynne Rienner, pp. 101–27.

Neumann, Roderick (1995), 'Local challenges to global agendas: conservation, economic liberalization and the pastoralists' rights movement in Tanzania', *Antipode*, **27** (4), 363–382.

Ng'ambi, Lois (2002), Interview with Samalila Ukhondo Waste Group worker Lois Ng'ambi, conducted by author and Wilma Nchito, December 23, 2002.

Ng'andu, N., G. Mulikita, S. Moyo, and B. Nkowane (1988), *A Socio-Economic and Public Health Profile of Kamanga Township, Lusaka*, Lusaka: Institute for Social and Economic Research.

Ngiloi, Anderson (1992), 'Study of solid waste management in the city of Dar es Salaam', unpublished MA thesis, University of Dar es Salaam, Dar es Salaam, Tanzania.

Nguluma, Huba (1997), 'Sustainable human settlement and environment in search of a strategy to improve urban infrastructure in Tanzania: a case of the Hanna Nassif Upgrading Project in Dar es Salaam', unpublished MSc thesis, Royal Institute of Technology, Stockholm, Sweden.

Ngware, Suleiman (2000a), 'Status of urban research and policies in Tanzania', in Ngware, S. and J. M. L. Kironde (eds), *Urbanising Tanzania: Issues, Initiatives, and Priorities*, Dar es Salaam: University of Dar es Salaam Press, pp. 7–21.

Ngware, Suleiman (2000b), 'Institutional capacity building: local self-governance under the multiparty system in Tanzania', in Ngware, S., L. Dzimbiri, and R. Ocharo (eds), *Multipartism and People's Participation*, Dar es Salaam: Tema Publishers, pp. 53–102.

Ngware, Suleiman, and J.M.L. Kironde (2000), *Urbanising Tanzania: Issues, Initiatives, and Priorities*, Dar es Salaam: Dar es Salaam University Press.

Nipashe (2003a), 'Wapinzani hawa vipofu – Keenja' ['Those in the opposition are blind – Keenja'], *Nipashe*, 21 December 2003, p. 1.

Nipashe (2003b), 'Waislamu kuishtaki serikali' ['Muslims are charging the government'], *Nipashe*, 31 May 2003, pp. 1 and 4.

Njovu, Alex (2004), 'UCZ cries foul over Oasis Forum', *Zambia Mail*, July 6 2004, p. 1.

Nkhuwa, D. (1996), 'Hydrogeological and engineering geological problems of urban development over karstified marble in Lusaka', *Mitt. Ing u. Hydrogeol.*, Heft, **63**, 251.

Nkhuwa, Daniel (2002), Personal communication with Daniel Nkhuwa, December 12 2002.

Nnkya, Tumsifu, James Materu, and Makame Muhajir (2000), *Report of the Evaluation Mission: Zanzibar Sustainable Program*, Dar es Salaam: United Nations Development Program and United Republic of Tanzania.

Novichi, M. (1992), 'Zambia: a lesson in democracy', *Africa Report*, November/December, 13–17.

Nsama, Thomas, and Kaswende, Kingsley (2004), 'Zambia's corruption fight affected by vested interest – DFID report', *Zambian Post*, July 14, 2004, p. 5.

Nyerere, Julius (1972), *Decentralisation*, Dar es Salaam: Government Printers.

Nyitambe, Joash (2003), Author interview with Joash Nyitambe, GIS coordinator, Kinondoni Municipal Council, June 18, 2003.

Odumosu, Tayo (2000), 'When refuse dumps become "mountains": responses to waste management in Metropolitan Lagos', in Tesi, M. (ed.), *The Environment and Development in Africa*, Lanham, Maryland: Lexington Books, pp. 225–247.

Okema, Mike (1995), 'CCM Council defeat could be pointer to general election', *The East African*, 12–18 June 1995, p. 6.

Ole-Mungaya, Michael (1990), 'A study on the development of traditional settlements in urban areas: Zanzibar case study', unpublished diploma project, Ardhi Institute (University College of Lands and Architectural Studies), Dar es Salaam, Tanzania.

Olofsson, Jenny, and Erika Sandow (2003), 'Towards a more sustainable city planning: a case study of Dar es Salaam, Tanzania', unpublished minor field study report, Umea University.

Ongala, Remmy (1989), *Songs for the Poor Man*, Beverly Hills, California: Real World Records.

Paoletto, Glen (1999), 'Urban governance in the new economy', in Inoguchi, Takashi, Edward Newman, and Glen Paoletto (eds), *Cities and the Environment*, Tokyo: United Nations University Press, pp. 297–325.

Parenteau, Rene (1996), 'Managing sustainable growth in Third World cities', in Keil, Roger, Gerda Wekerle, and David Bell (eds), *Local Places in the Age of the Global City*, Montreal: Black Rose Books, pp. 91–96.

Pearce, David, and R.K. Turner (1990), *Economics of Natural Resources and the Environment*, Baltimore, Maryland: Johns Hopkins University Press.

Peet, Richard, and Michael Watts (1993), 'Development theory and environment in an age of market triumphalism', *Economic Geography*, **69**.

Pelling, Mark (2003a), *The Vulnerability of Cities: Natural Disasters and Social Resilience*, London: Earthscan.

Pelling, Mark (2003b), 'Toward a political ecology of urban environmental risk', in Zimmerer, Karl, and Thomas Bassett (eds), *Political Ecology: an Integrative Approach to Geography and Environment-Development Studies*, New York: Guilford, pp. 73–93.

Pessel, Sigrid (2003), Author interview with Sigrid Pessel, Urban Management Advisor to the Zanzibar Sustainable Program, German Development Service, July 8 2003.

Petterson, Don (2002), *Revolution in Zanzibar: an American's Cold War Tale*, Boulder: Westview.

Phillips, S. (2002), 'Social capital, local networks and community development', in Rakodi, C. and T. Lloyd-Jones (eds.), *Urban Livelihoods: A People-Centred Approach to Reducing Poverty*, London: Earthscan, pp. 133–150.

Phiri, Brighton (2004), 'Mukuni seeks Mazoka's ban from Lwiindi's organization', *Zambian Post*, July 8 2004, p. 1.

Phiri, Patson (2004), 'De-registered forum wants dialogue with Sikapwasha', *Zambian Mail*, July 12 2004, p. 2.

Phiri, White (2003), Personal communication between author, Wilma Nchito, and White K. Phiri, Samalila Ukhondo Waste Group, Kamanga, January 9, 2003.

Potter, Robert, Tony Binns, Jennifer Elliott, and David Smith (1999), *Geographies of Development*, Harlow, Essex: Pearson Education.

Potts, Deborah (1994), 'Urban environmental controls and low income housing in Southern Africa', in Main, H. and S. Williams, (eds), *Environment and Housing in Third World Cities*, New York: Wiley, pp. 207–223.

Powell, Colin (2002), Speech of the US Secretary of State Colin Powell to the National Academy of Sciences meeting, 30 April 2002.

Power, Marcus (2003), *Rethinking Development Geographies*, London: Routledge.

Power, Marcus, and James Sidaway (2004), 'The degeneration of tropical geography,' *Annals of the Association of American Geographers*, **94** (3), 585–601.

Rakner, Lise (2003a), *Political and Economic Liberalisation in Zambia 1991–2001*, Uppsala, Sweden: Nordic Africa Institute.

Rakner, Lise (2003b), 'Analyzing political processes in the context of multiparty elections: Zambia 2001–2002', *African Social Research*, **45/46**, v–x.

Rakner, Lise, and Svasand, Lars (2003a), 'The quality of electoral processes: Zambian elections 1991–2001', *African Social Research*, **45/46**, 1–23.

Rakner, Lise, and Svasand, Lars (2003b), 'From dominant to competitive party system: the Zambian experience 1991–2001', *African Social Research*, **47/48**, 21–45.

Rakodi, Carole (1986a), 'Colonial urban planning in Northern Rhodesia and its legacy', *Third World Planning Review*, **8**, 193–218.

Rakodi, Carole (1986b), 'Housing in Lusaka: policies and progress', in Williams, G. (ed.), *Lusaka and its Environs*, Lusaka: Zambia Geographical Association, pp. 189–209.

Rakodi, Carole (1993), 'Planning for whom?', in Devas, Nick and Carole Rakodi (eds.), *Managing Fast Growing Cities*, New York: John Wiley, pp. 207–235.

Rakodi, Carole (1994), 'Zambia', in Tarver, John (ed.), *Urbanization in Africa: a Handbook*, pp. 342–363.

Rakodi, Carole (1997), *The Urban Challenge in Africa: Growth and Management of its Large Cities*, Tokyo: United Nations University Press.

Rakodi, Carole (2002a), 'Economic growth, wellbeing and governance in Africa's urban Sector', in Belshaw, Deryke and Ian Livingstone (eds), *Renewing Development in Sub-Saharan Africa: Policy, Performance and Prospects*, London: Routledge, pp. 309–327.

Rakodi, Carole (2002b), 'Economic development, urbanization and poverty', in Rakodi, C. and T. Lloyd-Jones (eds), *Urban Livelihoods: A People-Centred Approach to Reducing Poverty*, London: Earthscan, pp. 23–34.

Rapley, John (2004), *Globalization and Inequality: Neoliberalism's Downward Spiral*, Boulder, Colorado: Lynne Rienner.

Rasmussen, Lissi (1993), *Christian-Muslim Relations in Africa: the Cases of Northern Nigeria and Tanzania Compared*, London: British Academic Press.

Redclift, Michael (2002), 'Pathways to sustainability?', *Geography*, **87** (3), 189–96.

Reuben, Kabiti (2000), 'Community participation in domestic solid waste management in Lusaka', unpublished Bachelor of Science research project report, University of Zambia.

Riddell, Barry (1981), 'The geography of modernization in Africa: a re-examination', *Canadian Geographer*, **25**, 290–99.

Robertshaw, Rory, Antoinette Louw, and Anna Mtani (2003), *Crime in Dar es Salaam: Results of a City Victim Survey*, Pretoria, South Africa: Institute for Security Studies.

Robinson, Jenny (2002), 'Global and world cities: A view from off the map', *International Journal of Urban and Regional Research*, **26**, 3, 531–54.

Robinson, Jenny (2003), 'Postcolonialising geography: tactics and pitfalls', *Singapore Journal of Tropical Geography*, **24** (3), 273–89.

Rosander, Eva (1992), 'People's participation as rhetoric in Swedish development aid', in Dahl, G., and A. Rabo (eds), *Kam-Ap or Take-Off: Local Notions of Development*, Stockholm, Sweden: Stockholm Studies in Social Anthropology, pp. 35–68.

Rweyemamu, Robert (1995), 'CCM agreed to compete, but not to play by the rules', *The East African*, 12–18 June, 1995, p. 9.

Saasa, Oliver (with Jerker Carlsson) (2002), *Aid and Poverty Reduction in Zambia: Mission Unaccomplished*, Uppsala, Sweden: Nordic Africa Institute.

Sachs, Jeffrey (2000), 'The geography of economic development', Newport, RI: US Naval War College Jerome Levy Occasional Paper No. 1.

Sampson, Richard (1971), *So This Was Lusaakas: The Story of the Capital of Zambia*, 2nd edition, Lusaka: Multimedia.

Sampson, Richard (2002), *With Sword and Chain in Lusaka: A Londoner's Life in Zambia 1948–1972*, Victoria, BC: Trafford.

Sanderson, David, and Darren Hedley (2002), 'Strengthening urban livelihoods in Zambia – PUSH II and PROSPECT', in Rakodi, C. and T. Lloyd-Jones (eds.), *Urban Livelihoods: A People-Centred Approach to Reducing Poverty*, London: Earthscan, pp. 247–256.

Sandilands, Catriona (1996), 'The shaky ground of urban sustainability: a comment on ecopolitics and uncertainty', in Keil, Roger, Gerda Wekerle, and David Bell (eds), *Local Places in the Age of the Global City*, Montreal: Black Rose Books, pp. 125–130.

Sanya, Muhammad (2003), 'Maskini mji wetu mzuri?' ['Pity our good city?'], *Dira*, 18–24 July 2003, p. 9.

Sata, Joy (2004), 'Tribalists will not be spared, warns Veep', *Zambian Mail*, July 14, 2004, p. 1.

Satterthwaite, David (2002), 'Lessons from the experience of some urban poverty-reduction programmes', in Rakodi, C. and T. Lloyd-Jones (eds), *Urban Livelihoods: A People-Centred Approach to Reducing Poverty*, London: Earthscan, pp. 257–269.

Sawio, Camillus (1998), 'Managing urban agriculture in Dar es Salaam', unpublished consultancy for the Sustainable Dar es Salaam Project, Dar es Salaam, Tanzania.

Sawio, Camillus (1999), 'Urban agriculture development research training and network creation for Eastern and Southern Africa', paper presented at the International Research Workshop on African Environments: Technology, Modeling, and Political Ecology, Lawrence, Kansas, 8–11 September 1999.

Sawio, C., and C. Sokoni (1994), 'Human settlement and the environment: environmental policy considerations', in Mwandosya, M., M. Luhanga, and E. Mugurusi (eds), Environmental Protection and Sustainable Development, Dar es Salaam: Centre for Energy, Environment, Science, and Technology, pp. 113–140.

Schech, Susanne, and Jane Haggis (2000), *Culture and Development: a Critical Introduction*, Malden, Massachusetts: Blackwell.

Schlyter, Ann (1999), *Recycled Inequalities: Youth and Gender in George Compound, Zambia*. Uppsala: Nordiska Afrikainstitutet.

Schlyter, Ann (2002), *Empowered with Ownership: The Privatization of Housing in Lusaka, Zambia*, Roma, Lesotho: Institute of Southern African Studies.

Schneider, Jane, and Ida Susser (2003), *Wounded Cities: Destruction and Reconstruction in a Globalized World*. Oxford: Berg.

Schroeder, Richard (1999), 'Geographies of environmental intervention in Africa', *Progress in Human Geography*, **23** (3), 359–378.

Sellstrom, Tor (1999a), *Liberation in Southern Africa: Regional and Swedish Voices*, Uppsala, Sweden: Nordic Africa Institute.

Sellstrom, Tor (1999b), *Sweden and National Liberation in Southern Africa*, Uppsala, Sweden: Nordic Africa Institute.

Serikali ya Mapinduzi Zanzibar [Revolutionary Government of Zanzibar] (2003), 'Hotuba ya Waziri wa Maji, Umeme, Nishati na Ardhi, Mheshimiwa Burhan Saadat Haji, kuhusu makadirio ya mapato na matumizi kwa mwaka 2003/4 katika Baraza la Wawakilishi' [Speech by the Minister of Water, Electricity, Energy, and Lands, the Honorable Burhan Saadat Haji, concerning estimated revenues and expenditures for the budget year 2003/4], Zanzibar: Revolutionary Government of Zanzibar.

Seshamani, Venkatesh (2002), 'Employment and sustainable livelihoods in Zambia: issues, constraints, and prospects for improvement', *African Social Research*, **43/44**, 8–31.

Seymour, Tony, and unnamed sociology students (1978), 'Social survey in Ng'ombe squatter settlement', in van den Berg, Leo (ed.), *Hard Times in the City*, Lusaka: UNZA studies in Zambian Society, pp. 66–79.

Shani, Ali (2003), 'Hakuna sera za Manispaa' ['There are no policies of the Municipality'], *Dira*, 15–21 August 2003, p. 3.

Shao, Ibrahim, Angwara Kiwara, and George Makusi (1992), *Structural Adjustment in a Socialist Country: The Case of Tanzania*, Harare, Zimbabwe: Sapes Books.

Sheriff, Abdul (2002), 'The spatial dichotomy of Swahili towns: the case of Zanzibar in the nineteenth century', in Burton, Andrew (ed.), *The Urban*

Experience in Eastern Africa c. 1750–2000, Nairobi: British Institute in Eastern Africa, pp. 63–81.

Sheriff, Abdul, and Ed Ferguson (1991), *Zanzibar under Colonial Rule*, Athens, Ohio: Ohio University Press.

Shivji, Issa (1991), 'The politics of liberalisation in Tanzania: notes on the crisis of ideological hegemony', in Campbell, Horace, and Howard Stein (eds), *The IMF and Tanzania: The Dynamics of Liberalisation*, Harare, Zimbabwe: Sapes Books, pp. 67–85.

Shivji, Issa (1998), *Not Yet Democracy: Reforming Land Tenure in Tanzania*, London: International Institute for Environment and Development.

Shoo, Gideon, and Jesse Kwayu (1995), 'Ubinafsishaji ni wizi na udanganyifu –Mtei' ['Privatization is theft and fraud – Mtei'], *Rai*, 1–7 June, 1995, p. 5.

Shurmer-Smith, Pamela (2002), *Doing Cultural Geography*, London: Sage.

Sidaway, James (1998), 'The (geo)politics of regional integration: the example of the Southern African Development Community', *Environment and Planning D: Society and Space*, **16**, 549–576.

Sidaway, James (2000), 'Postcolonial geographies: an exploratory essay', *Progress in Human Geography*, **24** (4), 591–612.

Sikwibele, Anne (1996), 'Women, water supply, and sanitation problems in poor compounds of Lusaka', *Eastern Africa Social Science Research Review*, **12** (2), 37–52.

Simon, David (1995a), 'Debt, democracy and development: Sub-Saharan Africa in the 1990s', in Simon, D., W. van Spengen, C. Dixon, and A. Narman (eds), *Structurally Adjusted Africa*, London: Pluto Press, pp. 17–44.

Simon, David (1995b), 'The world city hypothesis: reflections from the periphery', in Knox, Paul and Peter Taylor (eds), *World Cities in a World System*, Cambridge: Cambridge University Press, pp. 132–155.

Simon, David (1997), 'Urbanization, globalization, and economic crisis in Africa', in Rakodi, Carole (ed.), *The Urban Challenge in Africa*, Tokyo: United Nations University Press, pp. 74–110.

Simon, David (1998), 'Rethinking (post)modernism, postcolonialism, and post-traditionalism: South-North perspectives', *Environment and Planning D: Society and Space*, **16**, 219–245.

Simon, David (1999a), 'Rethinking cities, sustainability, and development in Africa', in Zeleza, Paul and Ezekiel Kalipeni (eds), *Sacred Spaces and Public Quarrels: African Cultural and Economic Landscapes*, Trenton: Africa World Press, pp. 17–41.

Simon, David (1999b), 'Development revisited: thinking about, practicing and teaching development after the Cold War', in Simon, D. and A. Narman (eds), *Development as Theory and Practice: Current Perspectives on Development and Development Co-operation*, Harlow, Essex: Longman, pp. 17–54.

Simon, D., W. van Spengen, C. Dixon, and A. Narman (1995), *Structurally Adjusted Africa*, London: Pluto Press.

Simone, Abdoumaliq (1997), 'Urban development in South Africa: some critical issues from Johannesburg', in Burgess, Rod, Marisa Carmona, and Theo Kolstee (eds), *The Challenge of Sustainable Cities: Neoliberalism and*

Urban Strategies in Developing Countries, Atlantic Highlands, New Jersey: Zed Books, pp. 245–261.

Simone, Abdoumaliq (2001a), 'Straddling the divides: remaking associational life in the informal African city', *International Journal of Urban and Regional Research*, **25** (1), 102–117.

Simone, Abdoumaliq (2001b). 'On the worlding of African cities', *African Studies Review*, **44** (2), 15–42.

Simone, Abdoumaliq (2001c), 'Between ghetto and globe: remaking urban life in Africa', in Tostensen, Arne, Inge Tvedten, Mariken Vaa (eds), *Associational Life in African Cities: Popular Responses to the Urban Crisis*, Uppsala, Sweden: Nordic Africa Institute, pp. 46–63.

Simone, Abdoumaliq (2004), *For the City Yet to Come: Changing African Life in Four Cities*, Durham: Duke University Press.

Simwinga, Musonda, Macloud Nyirenda, Irene Mulundika, and Mina Brill (1997), 'Lusaka City Council community profiling survey of nine unplanned settlements', Lusaka: Lusaka City Council and Ireland Aid.

Sinyangwe, Binwell (2000), *A Cowrie of Hope*, Portsmouth, NH: Heinemann.

Soiri, Iina, and Pekka Peltola (1999), *Finland and National Liberation in Southern Africa*, Uppsala, Sweden: Nordic Africa Institute.

Soja, Edward (1968), *The Geography of Modernization in Kenya*, Syracuse: Syracuse University Press.

Soja, Edward (1989), *Postmodern Geographies: The Reassertion of Space in Critical Social Theory*, London: Verso.

Sommers, Marc (2001), *Fear in Bongoland: Burundi Refugees in Urban Tanzania*, New York: Berghahn Books.

Southern African Research and Documentation Center (SARDC) (1994), *State of the Environment in Southern Africa*. Harare: Southern African Research and Documentation Center.

Stamp, Dudley, and W. Morgan (1953), *Africa: A Study in Tropical Development*, New York: John Wiley and Sons.

Stone, Jeffrey (1978/79), 'The British Association essays on the human geography of Northern Rhodesia, 1931–35', *Zambian Geographical Journal*, **33–34**, 31–48.

Stren, R (1989a), 'The administration of urban services', in Stren, R. and R. White (eds), *African Cities in Crisis: Managing Rapid Urban Growth*, Boulder: Westview, pp. 37–67.

Stren, R. (1989b), 'Urban local government in Africa', in Stren, R. and R. White (eds), *African Cities in Crisis: Managing Rapid Urban Growth*, Boulder: Westview, pp. 20–36.

Sustainable Cities Program (2000), 'City experiences in improving the urban environment: a snapshot of an evaluation of six city initiatives in Africa, 1999', Nairobi: The United Nations Centre for Human Settlements and United Nations Environment Program Sustainable Cities Program, Working Paper no. 1.

Sustainable Cities Program (2001), 'Implementation and replication of the Sustainable Cities Programme process at city and national Level: case studies from nine Cities', Nairobi: The United Nations Centre for Human

Settlements and United Nations Environment Program Sustainable Cities Program, Working Paper no. 2.

Sustainable Cities Program (2004), Home page of the Sustainable Cities Program, at http://www.unchs.org/scp/Scphome.htm.

Swedish International Development Agency (SIDA) (2004), website of the Swedish International Development Agency, at http://www.sida.se.

Swilling, Mark (1997), *Governing Africa's Cities*, Johannesburg: Witwatersrand Press.

Swyngedouw, Erik (2003), 'Modernity and the production of the Spanish waterscape', in Zimmerer, Karl, and Thomas Bassett (eds), *Political Ecology: an Integrative Approach to Geography and Environment-Development Studies*, New York: Guilford, pp. 94–112.

Swyngedouw, Erik (2004), *Social Power and the Urbanization of Water: Flows of Power*, Oxford: Oxford University Press.

Sykes, Kleist, and Anna Mtani (2003), 'Governance challenges and coalition building among urban environmental stakeholders in Dar es Salaam, Tanzania', paper presented to the conference on the Urban Biosphere and Society Partnership of Cities, New York Academy of Sciences, 29 October 2003.

Tait, J. (1997), *From Self-Help Housing to Sustainable Settlement: Capitalist Development and Urban Planning in Lusaka, Zambia*, Brookfield, VT: Avebury.

Tandon, Yash (1999), *Globalization and Africa's Options*, Harare: International South Group Network.

Tanganyika Territory (1950), *Administration of Tanganyika Territory: Legislative and Administrative System*, Dar es Salaam: Government Printers.

Taylor, Mike, and Aarnes, Dag (2003), 'The 2001 Zambian elections in the context of economic decline', *African Social Research*, **45/46**, 151–196.

Tembo, Peter (2002), Interview by author and Wilma Nchito with Peter Tembo, Kwawama Waste Group, Ng'ombe, Lusaka, December 19, 2002.

Tembo, Peter (2003), Personal communication with author and Mweetwa Mudenda, January 6 2003.

Tembo, Trywell (1996), *The Road to Multi-Party Democracy in Zambia and its Consequences*, Livingstone: Sanisani Chemists.

Tibaijuka, Anna (2001), 'Preface', in 'United Nations Center for Human Settlements (Habitat), Tools to Support Participatory Urban Decision Making', Nairobi: UNCHS, p. i.

Times of Zambia (2002a), 'Dysentery hits displaced Ng'ombe families', *Times of Zambia*, December 25 2002, p. 3.

Times of Zambia (2002b), 'Lusaka Council complains', *Times of Zambia*, December 13 p. 1.

Times of Zambia (2003a), 'Ng'ombe squatters get land', *Times of Zambia*, January 1 2003, p. 1.

Times of Zambia (2003b), 'Keep Lusaka Clean Trust Fund to be launched', *Times of Zambia*, January 5 2003, p. 3.

Times of Zambia (2004), 'Change attitude, Lusaka mayor urges residents', *Times of Zambia*, April 29 2004, p. 1.

Toufiq, Salum (1998), 'Institutional responsibilities and capacity building for urban management in Zanzibar Municipality', unpublished proposition paper, Zanzibar Sustainable Program.

Tripp, Aili Marie (1997), *Changing the Rules: The Politics of Liberalization and the Urban Informal Economy in Tanzania*, Berkeley: University of California Press.

Tume ya Jiji la Dar es Salaam [Dar es Salaam City Commission] (2000), *Muhtasari wa Utekelezaji wa Majukumu ya Tume ya Jiji Kuanzia Julai 1996 hadi Novemba 1999* [*Summary of Implementation of Duties for the City Commission, from July 1996 to November 1999*], Dar es Salaam: Dar es Salaam City Commission.

United Nations Centre for Human Settlements (Habitat) (1992), *Managing the Sustainable Growth and Development of Dar es Salaam: Environmental Profile of the Metropolitan Area*, Dar es Salaam: UNDP and UNCHS.

United Nations Centre for Human Settlements (Habitat) (1998), *Financing Cities for Sustainable Development, with Special Reference to East Africa*, Nairobi: UNCHS.

United Nations Centre for Human Settlements (Habitat) (1999), *Establishing and Supporting a Working Group Process*, SCP Source Book Series, Nairobi: UNCHS and UNEP.

United Nations Centre for Human Settlements (Habitat) (2000), *Toolkit for Environmental Management Information Systems*, Nairobi: UNCHS.

United Nations Centre for Human Settlements (Habitat) (2001), 'Democratizing the urban environment agenda', *Habitat Debate*, 7 (1), 12 and 21.

United Nations Centre for Human Settlements (Habitat) (2001a), *Tools to Support Participatory Urban Decision Making*, Urban Governance Toolkit Series, Nairobi: UNCHS.

United Nations Centre for Human Settlements (Habitat) and United Nations Environment Program (1997), *Implementing the Urban Environment Agenda, Volumes 1–3: EPM Source Book*, Nairobi: UNCHS/UNEP.

UNDP and UNCHS (Habitat) (1993), 'Managing the sustainable growth and development of Dar es Salaam, Kunduchi Beach Hotel briefing seminar, 26th November, 1993: Summary of Proceedings (made available by Elias Chinamo, Director of Solid Waste Management, Dar es Salaam).

Urban Authorities Support Unit (UASU) (1998), *Strategic Urban Development Plan for Dar es Salaam*, Dar es Salaam: Sustainable Dar es Salaam Project.

Van Donge, Jan Kees (1998), 'Reflections on donors, opposition and popular will in the 1996 Zambian general elections', *Journal of Modern African Studies*, 36 (1), 71–99.

Wangwe, Samuel, Haji Semboja, and Paula Timbandebage (1998), *Transitional Economic Policy and Policy Options in Tanzania*, Dar es Salaam: Mkuki na Nyota Publishers.

Watts, Michael (1987), 'Powers of production: geographers among the peasants', *Environment and Planning D: Society and Space*, 5 (2), 215–30.

Watts, Michael (1993a), 'Development I: power, knowledge, discursive practice', *Progress in Human Geography*, 17 (2), 257–72.

Watts, Michael (1993b), 'The geography of post-colonial Africa: space, place, and development in Sub-Saharan Africa (1960–93)', *Singapore Journal of Tropical Geography*, **14** (2), 173–90.

Watts, Michael (1994), 'Development II: the privatization of everything?', *Progress in Human Geography*, **18** (3), 371–84.

Watts, Michael (1997), 'Black gold, white heat: state violence, local resistance and the national question in Nigeria', in Pile, Steve and Michael Keith (eds.), *Geographies of Resistance*, London: Routledge, pp. 33–67.

Watts, Michael (2003), 'Development and governmentality', *Singapore Journal of Tropical Geography*, **24** (1), 6–34.

Watts, Michael (2004), 'Antinomies of community: some thoughts on geography, resources and empire', *Transactions, Institute of British Geographers*, **29** (2), 195–216.

Williams, Geoffrey (1986), 'The early years of the township', in Williams, G. (ed.), *Lusaka and its Environs*, Lusaka: Zambia Geographical Association, pp. 71–94.

Wolgin, Jerome (1999), 'The foundations of sustainable development in Africa: ten trends and one conundrum', in James, Valentine (ed.), *Sustainable Development in Africa: Prospects for the 21st Century*, San Francisco: International Scholars Publications, pp. 9–22.

Wood, Adrian, G.P. Banda, and D.C. Mundende (1986), 'The population of Lusaka', in Williams, G. (ed.), *Lusaka and its Environs*, Lusaka: Zambia Geographical Association, pp. 164–88.

World Bank (2001), *Tanzania at the Turn of the Century: From Reforms to Sustained Growth and Poverty Reduction*, Washington, DC: The World Bank.

World Bank (2002), *Upgrading of Low Income Settlements Country Assessment Report: Zambia*, Washington, DC: The World Bank.

Wunsch, James, and Dele Olowu (1990), *The Failure of the Centralized State: Institutions and Self-Governance in Africa*, Boulder: Westview.

Yeoh, Brenda (2001), 'Postcolonial cities', *Progress in Human Geography*, **25** (3), 456–468.

Yhdego, Michael (1995), 'Waste management – case study Dar es Salaam', in Njau, G. and E. Mugurusi (eds), *Towards Sustainable Environment in Tanzania*, Dar es Salaam: Ministry of Natural Resources, Tourism, and Environment, pp. 117–125.

Yussuf, Issa (1995), 'Fedha za SMZ kuhamishwa BoT' ['Treasury of the Revolutionary Government of Zanzibar to be moved to the Bank of Tanzania'], *Nuru*, 3–9 July 1995, p. 1.

Zambian Post, 'Harare mayor released from jail', *Zambian Post*, January 14, 2003, p. 4.

Zambian Post, 'Levy, change your attitude', *Zambian Post*, July 9, 2004, p. 10.

Zanzibar Leo (2003a), 'Kampeni maalum ya usafi kuanza leo' ['Formal cleanup campaign starts today'], *Zanzibar Leo*, 6 September 2003, p. 1.

Zanzibar Leo (2003b), 'Manispaa kuandaa sheria kali kudhibiti ujenzi holela' ['Municipality prepares harsh law to prohibit haphazard construction'], *Zanzibar Leo*, 15 September 2003, p. 1.

Zanzibar Leo (2003c), 'Ubinafsishaji wa huduma za uzoaji taka ufikiriwe' ['Privatization of waste collection considered'], *Zanzibar Leo*, 7 September 2003, p. 1.

Zanzibar, Revolutionary Government of (1997), *Project Support Implementation Arrangements for the Zanzibar Sustainable Project*, Zanzibar, Tanzania: Revolutionary Government of Zanzibar and United Nations Development Program.

Zanzibar Sustainable Program (1998), *Environmental Profile of Zanzibar Municipality*. Zanzibar: Zanzibar Sustainable Program.

Zanzibar Sustainable Program (2000), 'Working group report: municipal financing, Zanzibar Municipality', Zanzibar: Zanzibar Sustainable Program.

Zanzibar Sustainable Program (2002), 'Towards a sustainable urban environment', Zanzibar: Zanzibar Sustainable Program.

Zanzibar Sustainable Program (2003), *Jarida*, 1st issue of a newsletter, July 2003, Zanzibar: Mradi wa Uendelevu wa Mji wa Zbar (ZSP).

Zeleza, Paul (1999), 'The spatial economy of structural adjustment in African cities', in Zeleza, Paul, and Ezekiel Kalipeni (eds.), *Sacred Spaces and Public Quarrels: African Cultural and Economic Landscapes*, Trenton, New Jersey: Africa World Press, pp. 43–71.

Zimmerer, Karl, and Thomas Bassett (2003), *Political Ecology: An Integrative Approach to Geography and Environment-Development Studies*, New York: Guilford.

Zukas, Simon (2002), *To Exile and Back*, Lusaka: Bookworld Publishers.

Index